Communications
in Computer and Information Science 2020

Editorial Board Members

Rationale
The CCIS series is devoted to the publication of proceedings of computer science conferences. Its aim is to efficiently disseminate original research results in informatics in printed and electronic form. While the focus is on publication of peer-reviewed full papers presenting mature work, inclusion of reviewed short papers reporting on work in progress is welcome, too. Besides globally relevant meetings with internationally representative program committees guaranteeing a strict peer-reviewing and paper selection process, conferences run by societies or of high regional or national relevance are also considered for publication.

Topics
The topical scope of CCIS spans the entire spectrum of informatics ranging from foundational topics in the theory of computing to information and communications science and technology and a broad variety of interdisciplinary application fields.

Information for Volume Editors and Authors
Publication in CCIS is free of charge. No royalties are paid, however, we offer registered conference participants temporary free access to the online version of the conference proceedings on SpringerLink (http://link.springer.com) by means of an http referrer from the conference website and/or a number of complimentary printed copies, as specified in the official acceptance email of the event.

CCIS proceedings can be published in time for distribution at conferences or as post-proceedings, and delivered in the form of printed books and/or electronically as USBs and/or e-content licenses for accessing proceedings at SpringerLink. Furthermore, CCIS proceedings are included in the CCIS electronic book series hosted in the SpringerLink digital library at http://link.springer.com/bookseries/7899. Conferences publishing in CCIS are allowed to use Online Conference Service (OCS) for managing the whole proceedings lifecycle (from submission and reviewing to preparing for publication) free of charge.

Publication process
The language of publication is exclusively English. Authors publishing in CCIS have to sign the Springer CCIS copyright transfer form, however, they are free to use their material published in CCIS for substantially changed, more elaborate subsequent publications elsewhere. For the preparation of the camera-ready papers/files, authors have to strictly adhere to the Springer CCIS Authors' Instructions and are strongly encouraged to use the CCIS LaTeX style files or templates.

Abstracting/Indexing
CCIS is abstracted/indexed in DBLP, Google Scholar, EI-Compendex, Mathematical Reviews, SCImago, Scopus. CCIS volumes are also submitted for the inclusion in ISI Proceedings.

How to start
To start the evaluation of your proposal for inclusion in the CCIS series, please send an e-mail to ccis@springer.com.

Jose M. Juarez · Carlos Fernandez-Llatas ·
Concha Bielza · Owen Johnson · Primoz Kocbek ·
Pedro Larrañaga · Niels Martin ·
Jorge Munoz-Gama · Gregor Štiglic ·
Marcos Sepulveda · Alfredo Vellido
Editors

Explainable Artificial Intelligence and Process Mining Applications for Healthcare

Third International Workshop, XAI-Healthcare 2023
and First International Workshop, PM4H 2023
Portoroz, Slovenia, June 15, 2023
Proceedings

Springer

Editors

Jose M. Juarez (ID)
University of Murcia
Murcia, Spain

Concha Bielza (ID)
Universidad Politécnica de Madrid
Boadilla del Monte, Madrid, Spain

Primoz Kocbek (ID)
University of Maribor
Maribor, Slovenia

Niels Martin (ID)
Hasselt University
Diepenbeek, Belgium

Gregor Štiglic (ID)
University of Maribor
Maribor, Slovenia

Alfredo Vellido (ID)
Universitat Politècnica de Catalunya
Barcelona, Spain

Carlos Fernandez-Llatas (ID)
Universitat Politècnica de València
Valencia, Spain

Owen Johnson (ID)
University of Leeds
Leeds, UK

Pedro Larrañaga (ID)
Universidad Politécnica de Madrid
Boadilla del Monte, Spain

Jorge Munoz-Gama (ID)
Pontificia Universidad Católica de Chile
Macul, Chile

Marcos Sepulveda (ID)
Pontificia Universidad Católica de Chile
Santiago, Chile

ISSN 1865-0929   ISSN 1865-0937 (electronic)
Communications in Computer and Information Science
ISBN 978-3-031-54302-9   ISBN 978-3-031-54303-6 (eBook)
https://doi.org/10.1007/978-3-031-54303-6

This Springer imprint is published by the registered company Springer Nature Switzerland AG
The registered company address is: Gewerbestrasse 11, 6330 Cham, Switzerland

Paper in this product is recyclable.

Preface

The Artificial Intelligence in Medicine (AIME) society was established in 1986 with the main goals of fostering fundamental and applied research in the application of Artificial Intelligence (AI) techniques to medical care and medical research and providing a forum for discussing any progress made. For this purpose, a series of AIME conferences have been organized on a biennial basis since 1987.

The 21st edition of the AIME conference was held in Portoroz, Slovenia, in June 2023. Five workshops were organized in conjunction with the AIME 2023 main conference. This volume contains a selection of the best papers presented at the 3rd International Workshop on eXplainable Artificial Intelligence in Healthcare (XAI-Healthcare 2023) and the 1st International Workshop on Process Mining Applications for Healthcare (PM4H 2023).

XAI-Healthcare 2023 provided a place for intensive discussion on all aspects of eXplainable Artificial Intelligence (XAI) in the medical and healthcare field (https://www.um.es/medailab/events/XAI-Healthcare/). This resulted in cross-fertilization among research on Machine Learning, Decision Support Systems, Natural Language, Human-Computer Interaction, and Healthcare sciences. This meeting also provided attendees with an opportunity to learn more on the progress of XAI in healthcare and to share their own perspectives.

The XAI-Healthcare 2023 workshop received 11 submissions. All papers were carefully peer reviewed (single-blind review) by at least 2 experts from the Program Committee. The reviewers judged the overall quality, the novelty, and the relevance to the XAI-Healthcare 2023 workshop. As a result, 7 papers were finally accepted. Each paper was presented in a 10-minute oral presentation. The workshop also had the privilege of hosting the keynote speaker Mihaela van der Schaar, professor of Machine Learning, Artificial Intelligence and Medicine at the University of Cambridge and fellow at The Alan Turing Institute in London. She discussed and shared her experience about the new frontiers in Machine Learning interpretability in the healthcare field.

The 2023 International Workshop on Process Mining Applications for Healthcare (PM4H 2023) offered a distinguished platform for interdisciplinary researchers and practitioners to share insights and ideas regarding data-driven process analysis techniques in healthcare. Within the Artificial Intelligence in medicine community, PM4H 2023 fostered discussions on a wide array of topics, spanning from tailored process mining techniques for healthcare processes to practical challenges associated with implementing PM4H methodologies in healthcare organizations. PM4H 2023 was organized by the Process Oriented Data Science Alliance (https://pods4h.com/alliance) in cooperation with the IEEE Task Force on Process Mining (https://www.tf-pm.org/), having the same academic quality standards and Program Committee as the PODS4H workshop (https://pods4h.com/) that this year reached its 6th edition.

The PM4H 2023 workshop garnered 17 submissions, each undergoing meticulous single-blind peer review by a minimum of 3 experts from the Program Committee.

Reviewers evaluated the papers based on overall quality, novelty, and their applicability to real healthcare scenarios. Ultimately, we approved 6 papers for 15-minute oral presentations and 2 for poster sessions. Among these, 5 papers were chosen for publication in the proceedings.

We finally would like to express our gratitude to the AIME 2023 organization, to the invited keynote speakers for their participation, and to all the members of the program committees for their invaluable support in making the XAI-Healthcare and PM4H workshops a success. We would also like to extend our gratitude to the EasyChair conference management system for the support provided, and to Springer for the trust placed in this endeavor.

We hope that you will find our selection of papers from the XAI-Healthcare and PM4H 2023 workshops included in this proceedings volume interesting and stimulating.

December 2023

Jose M. Juarez
Carlos Fernandez-Llatas
Concha Bielza
Owen Johnson
Primoz Kocbek
Pedro Larrañaga
Niels Martin
Jorge Munoz-Gama
Gregor Štiglic
Marcos Sepúlveda
Alfredo Vellido

Organization

Chairs of International Workshop on Explainable Artificial Intelligence in Healthcare

Concha Bielza	Universidad Politécnica de Madrid, Spain
Jose M. Juarez	University of Murcia, Spain
Primoz Kocbek	University of Maribor, Slovenia
Pedro Larrañaga	Universidad Politécnica de Madrid, Spain
Gregor Štiglic	University of Maribor, Slovenia
Alfredo Vellido	Universitat Politècnica de Catalunya (UPC BarcelonaTech), Spain

Chairs of International Workshop on Process Mining applications for Healthcare

Carlos Fernandez-Llatas	Universitat Politècnica de València, Spain
Niels Martin	University of Hasselt, Belgium
Owen Johnson	University of Leeds, UK
Marcos Sepúlveda	Pontificia Universidad Católica de Chile, Chile
Jorge Munoz-Gama	Pontificia Universidad Católica de Chile, Chile

Program Committee of International Workshop on Explainable Artificial Intelligence in Healthcare

Alejandro Rodriguez	Universidad Politécnica de Madrid, Spain
Bernardo Canovas-Segura	University of Murcia, Spain
Carlo Combi	University of Verona, Italy
Caroline König	Universitat Politècnica de Catalunya, Spain
Huang Zhengxing	Zhejiang University, China
Jean-Baptiste Lamy	Université Sorbonne Paris Nord, France
Jose M. Alonso	Universidad de Santiago de Compostela, Spain
Lluis Belanche	Universitat Politècnica de Catalunya, Spain
Milos Hauskrecht	University of Pittsburgh, USA
Nava Tintarev	Maastricht University, The Netherlands
Paulo Felix	Universidad de Santiago de Compostela, Spain
Pedro Cabalar	University of Coruna, Spain

Przemyslaw Biecek	Warsaw University of Technology, Poland
Riccardo Bellazzi	University of Pavia, Italy
Shashikumar Supreeth	University of California San Diego, USA
Pham Thai-Hoang	Ohio State University, USA
Zhe He	Florida State University, USA

Program Committee of International Workshop on Process Mining Applications for Healthcare

Davide Aloini	University of Pisa, Italy
Iris Beerepoot	Utrecht University, The Netherlands
Elisabetta Benevento	University of Pisa, Italy
Andrea Burattin	Technical University of Denmark, Denmark
Arianna Dagliatti	University of Pavia, Italy
Benjamin Dalmas	Centre de Recherche Informatique de Montréal, Canada
Rene de la Fuente	Pontificia Universidad Católica de Chile, Chile
Kerstin Denecke	Bern University of Applied Sciences, Switzerland
Hans Eguia	Universitat Oberta de Cataluña, Spain
Carlos Fernandez-Llatas	Universitat Politècnica de València, Spain
Roberto Gatta	Università Cattolica S. Cuore, Italy
Josha Grueger	University of Trier, Germany
Emmanuel Helm	University of Applied Sciences Upper Austria, Austria
Gema Ibanez-Sanchez	Universitat Politècnica de València, Spain
Owen Johnson	University of Leeds, UK
Luis Marco	Norwegian Centre for E-Health Research, Norway
Mar Marcos	Universitat Jaume I, Spain
Niels Martin	University of Hasselt, Belgium
Begoña Martinez	Universitat Jaume I, Spain
Renata Medeiros de Carvalho	Eindhoven University of Technology, The Netherlands
Jorge Munoz-Gama	Pontificia Universidad Católica de Chile, Chile
Simon Poon	University of Sydney, Australia
Luise Pufahl	TU Berlin, Germany
Hajo Reijers	Utrecht University, The Netherlands
Octavio Ribera-Romero	Universidad de Sevilla, Spain
Eric Rojas	Universidad Católica de Chile, Chile
Massimiliano Ronzani	Fundazione Bruno Kessler, Italy
Lucia Sacchi	University of Pavia, Italy

Contents

International Workshop on Explainable Artificial Intelligence in Healthcare

Unlocking the Power of Explainability in Ranking Systems: A Visual Analytics Approach with XAI Techniques

Mozhgan Salimiparasa$^{(\boxtimes)}$ ⬤, Kamran Sedig, and Daniel Lizotte

University of Western Ontario, London, ON N6A 3K7, Canada
msalimip@uwo.ca

Abstract. Ranking systems are widely used in various domains, including healthcare, to support decision making. However, understanding how these systems generate rankings can be challenging for users. In this paper, we present a visual analytic tool that combines XAI methods and interactive visualizations to explain ranking systems. Our tool provides users with a better understanding of how these systems work by using customized counterfactual explanations and feature importance visualizations. Unlike traditional counterfactual explanations that identify the minimum changes required to change class prediction, our tool considers a dynamic threshold that is set by other items in the list of items to be ranked. This threshold determines what changes are required for a specific item to rank lower or higher in the list. Our feature importance visualization shows the impact of each feature on the prediction, providing users with insights into how the system generates rankings. To demonstrate the effectiveness of our tool, we applied it to triage patients and rank them for admission to the ICU based on their severity. We demonstrated that our tool can provide clinicians with a better understanding of the ranking system and helped them make informed decisions about patient care. Our tool can also be applied to other ranking systems in healthcare and other domains, providing users with a transparent and understandable system for ranking-based decision support.

Keywords: Counterfactual Explanation · Ranking Systems · Visual Analytics Tool

1 Introduction

Ranking systems have become prevalent in various domains, including e-commerce [1], social media [2], and human resources [3,4]. These systems are often complex, with many input variables and black box algorithms, making it challenging for users to understand how they generate rankings. The lack of transparency and accountability in such systems presents a significant challenge, which may lead to user mistrust and reluctance in adopting them. Thus, there has been a growing demand to provide users with a deeper understanding of

© The Author(s), under exclusive license to Springer Nature Switzerland AG 2024
J. M. Juarez et al. (Eds.): XAI-Healthcare/PM4H 2023, CCIS 2020, pp. 3–13, 2024.
https://doi.org/10.1007/978-3-031-54303-6_1

these black box models, giving rise to the development of Explainable Artificial Intelligence (XAI) methods.

XAI methods are designed to enhance the transparency and interpretability of artificial intelligence systems for human comprehension. These methods aim to alleviate the difficulties in understanding how the ranking system generates its output by providing users with insights into its workings. By doing so, XAI methods improve transparency and user trust, leading to greater accountability and effective utilization of the system [5, 6].

In this paper, we present a visual analytics tool that uses XAI methods and feature importance visualization to explain ranking systems. Our tool provides users with a better understanding of how these systems generate rankings by identifying the impact of each feature on the output. We also introduce a customized counterfactual explanation method that considers a dynamic threshold for ranking items instead of the traditional approach of identifying the minimum changes required to change class prediction. To demonstrate the effectiveness of our tool, we apply it to a healthcare scenario of triage patients and ranking them for admission to the ICU based on their severity. Our tool can provide clinicians with a transparent and understandable system for decision-making, improving their confidence in the system and enabling them to make informed decisions about patient care.

We believe our tool can be applied to other ranking systems in various domains, providing users with a transparent and understandable system for ranking-based decision support. The rest of the paper is structured as follows: Sect. 2 reviews related work in explainable AI and ranking systems. Section 3 describes our methodology. Section 4 presents the experimental results of our tool applied to the healthcare scenario. Section 5 discusses the strengths and limitations of our approach, and Sect. 6 concludes the paper.

2   Related Work

Recently, the lack of interpretability of existing ranking techniques has received attention, leading to the development of XAI methods and visualization frameworks to aid in the interpretation and analysis of ranking models. For instance, Srvis proposed by Di Weng et al. integrate spatial contexts with rankings through scalable visualizations and flexible spatial filtering and comparative analysis to support decision-making for large-scale spatial alternatives like selecting store locations [7].

RankViz, on the other hand, supports the analysis and interpretation of learning-to-rank (LtR) models by providing visualizations that give information on important data features and enable comparison of element positions to aid in understanding and creation of rankings [8]. LineUp is a bar chart-based technique that ranks items based on multiple attributes with different scales and semantics, allowing interactive combination and refinement of parameters to explore changes in rankings and enable comparison of multiple rankings on the same set of items [9]. Meanwhile, uRank is a tool that provides views summarizing the contents of a recommendation set and interactive methods to convey users' interests through a recommendation ranking, enabling users to understand, refine, and reorganize documents as information needs evolve [10].

Hadis, et al. propose a hierarchical ranking explanation framework that uses a proper neighborhood construction approach to capture local explanations for competitive rankings, exploring various explanation techniques to identify the local contribution of ranking indicators based on an instance's position in the ranking and the size of the neighborhood [11]. Finally, Honglei, et al. introduce the use of generalized additive models (GAMs) for ranking tasks, instantiated using neural networks, demonstrating that their approach outperforms traditional GAMs while maintaining similar interpretability, offering promise for the development of intrinsically interpretable ranking models [12].

3   Methodology

Our visual analytics tool combines two key strategies for explaining rankings: XAI-derived explanations and interactive visualizations. The XAI strategies we use are counterfactual explanations and feature importance, and we have proposed a customized algorithm to contextualize counterfactual explanations for use in ranking systems. For feature importance, we use the Shapley value method. The interactive visualization strategies we use contain two primary sub-visualizations, namely the ranking list and the what-if panel. These visual representations serve as a bridge between the explanations generated by the XAI module and the users' comprehension, allowing them to explore the ranking system and gain a deeper understanding of how it operates.

3.1   XAI

Feature Importance. To determine the importance of features, we employed the SHAP method. SHAP (SHapley Additive exPlanations) is an XAI method that explains how machine learning models make predictions. It assesses the importance of each feature in the input data by determining how much each feature contributes to the model's output. The basis of SHAP values is the concept of Shapley values from cooperative game theory, which measures the contribution of each player in a game. In the context of machine learning, SHAP values measure the contribution of each feature to the prediction made by the model. They are computed by examining all possible combinations of features and calculating the amount each feature contributes to the prediction in comparison to its contribution in the absence of other features. The SHAP approach considers the marginal contribution of each feature in the prediction, which is the difference between the prediction made with and without the feature [13]. By assuming that the prediction is a weighted sum of feature values, SHAP is able to identify the weight of each feature based on its contribution to the prediction, thus allowing it to capture feature interactions. We selected this approach as it offers an understandable interpretation of the model's prediction, and its ability to capture interactions between features. Furthermore, SHAP is a model-agnostic method. It can be used with various types of machine learning models, including neural networks, tree-based models, and linear models, making them a flexible tool for analyzing feature importance in a broad range of machine learning applications.

Counterfactual Explanations. A counterfactual explanation describes the smallest change to the feature values of an example that results in the model making a meaningful change in output [14,15]. This is typically defined as a change in the predicted class (for classification), or the prediction reaching a pre-specified threshold (for probability or regression outputs). Counterfactual examples are explored as a way to investigate how tweaking feature values affects the output of the model.

Different ways of generating counterfactual examples have been studied [16–18]. The most popular approaches use optimization. Wachter et al. use an optimization algorithm to generate a new input that is the closest sample to the data instance but makes the model produce a different prediction [19]. Russell et al. used the idea of generative models to synthesize new input instances that are close to the original data but output models with different results [20]. Dandl et al. use a gradient-based optimization algorithm to find counterfactual examples [21]. Another approach to generating counterfactual examples is greedy search. In this approach, feature values are modified iteratively until the model prediction changes. Yang et al. used this approach to find counterfactual examples [22]. Prior works have focused on generating counterfactual examples using different approaches with the aim of explaining the minimum changes required to alter class predictions. In our approach, we modify the traditional greedy algorithm to address the specific needs of ranking problems. To achieve this, we generate counterfactual examples that take into account the relationship between items in a ranked list, providing a better understanding of the changes required to reach a desired rank.

Our proposed method uses a greedy algorithm to find counterfactual explanations that are applicable to ranked model outputs, which are a collection of outputs for a set of test data that have been sorted according to the output of a probabilistic classification model. The goal of our algorithm is to determine the minimum changes required for a data instance to achieve a different rank. For this purpose, we use a greedy approach given in Algorithm 1. The algorithm takes as input: *Mlmodel* - the machine learning model that produced the ranking; *dataInstance* - a specific data instance; and *RChange* - a desired rank change.

The algorithm first finds the most important features $\mathcal{F}^*$ sorted in descending order for the given *dataInstance* according to Shapley values. Then the algorithm initializes the set $\mathcal{F}$ with f^* (the first important feature in $\mathcal{F}^*$) and generates *InteractionList* which is a list of features that have the most interaction with this feature. These are the features whose values have the biggest impact on the relationship between the f^* feature and the outcome; they are available in standard implementations of XGBoost and can be computed for other models as well. To generate counterfactual examples, the approach varies the feature values in the subset *Fsub*, which is obtained by adding one feature at a time from the feature list $\mathcal{F}$). The approach changes the values of these features for *dataInstance* along a specified range using a grid search, and observing the corresponding model outputs while holding all other features constant. It computes the rank assigned by the model for each new potential counterfactual example and compares it to the rank of the original data instance.

Given that the proposed approach aims to achieve the desired ranking with the minimum number of changes possible, if a single important feature modification is sufficient to meet the target ranking, then modifying other features can be avoided. However, if changing the value of one of the features in *Fsub* alone is not able to change the rank of *dataInstance*, the algorithm considers changing multiple important features together to see if a change in rank can be produced. After each modification to the feature(s), the algorithm replaces the feature value with the modified value and evaluates the *MLmodel*'s output. The premise is that by simultaneously changing important features together with their most strongly interacting features, the algorithm has the best chance of being able to identify a counterfactual example whose rank is at least *RChange* away from the rank of *dataInstance*. The algorithm iteratively adds features from the list of most important features and their corresponding interaction list and performs a grid search after each is added. This process is repeated until a change in the feature values results in a counterfactual example whose rank is different from that of *dataInstance* by the desired amount (Rchange), at which point the algorithm will stop and return the new counterfactual example.

The objective of our approach is to identify a counterfactual example that involves the minimum possible number of feature modifications. To achieve this, we begin with the most important feature and add its interacting feature one at a time. If this process fails to yield a suitable counterfactual example, it will add another important feature (f^*) from $\mathcal{F}^*$ along with its interacting features to $\mathcal{F}$ and repeat the algorithm.

3.2 Interactive Visualization

Our interactive visualization is designed to provide users with a transparent and understandable system for decision support. Two primary sub-visualizations are employed, namely the ranking list and the what-if panel. These visualizations map the explanations produced in the XAI module into visual representations, allowing users to explore the ranking system and understand how it generates a ranking.

Interactive Ranking List. An interactive ranking list was developed based on the concept of semantic zoom to allow for a detailed exploration of each item locally and a comparison between items globally. Semantic zoom is a technique that provides users with distinct representations of data as they zoom in or out, with the aim of enhancing the overall understanding of the underlying

---

Algorithm 1. Generating Counterfactual Explanations for Ranking Models Using a Greedy Algorithm

---

1: **procedure** COUNTERFACTUALRANKING(Mlmodel, dataInstance, RChange)
2:     *InstanceRank* ← Mlmodel(dataInstance)
3:     **if** RChange is not provided **then**
4:         RChange ← 1
5:     $\mathcal{F}^*$ ← FindListOfMostImportantFeatures(Mlmodel, dataInstance)
6:     $\mathcal{F}$ ← *EmptyList*
7:     *Fsub* ← *EmptyList*
8:     **for** f^* in $\mathcal{F}^*$ **do**
9:         *InteractionList* ← GetFeaturesWithHighestInteractionWith(f^*, Mlmodel, dataInstance)
10:         $\mathcal{F}$ ← *append*($\mathcal{F}, f^*, InteractionList$)
11:         **for** f in $\mathcal{F}$ **do**
12:             *Fsub* ← *append*(*Fsub*, f)
13:             Δ_f ← {MinObserved(f),0,MaxObserved(f)}
14:             **for** δ in $\prod_{f \in Fsub} \Delta_f$ **do**   ▷ δ is a change to every feature (there are $3^{|Fsub|}$). This exhaustive search can be replaced with a heuristic search.
15:                 *newInput* ← ReplaceSelectedFeatureValues(dataInstance, δ)
16:                 *newRank* ← Mlmodel(*newInput*)
17:                 **if** *newRank* ≥ *InstanceRank* − RChange **or** *newRank* ≤ *InstanceRank* + RChange **then**
18:                     **return** *newInput*
19:     **return** "No feasible changes achieve the desired ranking."

---

semantic structure [23]. When an item in the ranking list is expanded, a treemap of its attributes is displayed, presenting the details of each item with different categories, if applicable. A treemap is a visualization technique that organizes hierarchical data into a set of nested rectangles. Each rectangle's size is proportional to a quantitative variable, and additional information about specific categories or variables can be conveyed through colors [24]. In our visualization the size of the boxes in the treemap represents the importance of each category and its contribution to the overall ranking, enabling users to identify the primary feature category responsible for the rank. Different colors are used for different feature categories. Additionally, each box is further divided into subcategories that specify the precise feature and its importance for the item's ranking through a prediction model. By using the treemap and its interactive features, such as zooming in and out, users can explore the categories, their features and values, and their contributions to each item in the ranking list. Semantic zoom facilitates the exploration of information at different levels of granularity. As an illustration, users can begin with an overview of the ranking list, subsequently zoom in to examine the top-ranked items more closely, and then zoom even further to inspect the particular attributes that contributed to their respective rankings. This feature enables users to navigate and comprehend large volumes of information without feeling overwhelmed.

What-If Panel. We developed a *what-if* that allows for the exploration of counterfactual scenarios and the necessary adjustments to feature values to alter an item's ranking. This panel contains histograms of the top 5 global feature importance values, offering users a comprehensive overview of the most critical features across the entire dataset. Users can select an item from the ranking list and set their preferred rank for that item. Our customized counterfactual explanation algorithm then identifies the minimum changes required to achieve the desired rank, which the system presents to the user. The system displays the value of the most important feature of the selected item with a blue line on the histogram, along with an indicator of its rank. This feature enables users to compare the feature values of different items with various rankings. Additionally, there are red and black dotted lines on the histogram, which represent the feature value required for the item to rank one rank above or below its current rank, respectively. In some cases, a feature may lack either a black or red line, indicating that changing that feature alone cannot alter the item's rank, and modifying more features is required. Furthermore, our system includes a box for the desired rank, which users can use to explore the required changes to the feature values for a specific item to reach the desired rank. Our system considers both lower and higher desired ranks. Users can also use radio buttons to specify which features they want to modify. This panel provides users with an interactive tool to better comprehend how the ranking system works. By displaying the top global features and allowing users to modify the feature values and observe their impact on the ranking, our system facilitates a more nuanced and in-depth understanding of the ranking system, which can aid users in making more informed decisions.

The rest of the paper will describe the experimental results of our tool applied to a healthcare scenario and discuss the strengths and limitations of our approach.

4   Case Study: Explaining Triaging Patients to Be Admitted to ICU

In order to evaluate the effectiveness of our proposed method, we conducted a case study focusing on ranking patients for admission to the ICU. For this purpose, we used a dataset from Sírio Libanês Hospital, which contains patient demographic information, previous disease groupings, blood results, vital signs, and blood gases [25]. The dataset includes labels indicating whether a patient was admitted to the ICU or not. We used this label to train an XGBoost model to predict patient admission probability. We then used these prediction probabilities to rank a test set of patients for triage purposes. After ranking the patients in the test set, our proposed method was applied to give a better understanding of how patients were ranked.

Figure 1 provides an overview of the entire visual analytics tool including the interactive ranking list on the left and the what-if panel on the right. The interactive list of ranked patients, as demonstrated in Fig. 1a, can be expanded to reveal a treemap showing the importance of the features used in the ranking. The treemap is designed to enable zooming in and out to explore the ranking importance of each category while hovering over each box generates a tooltip displaying the feature value. The size of each box corresponds to the importance of the feature. For instance, in Fig. 1a the blood test has the most impact on the ranking, and among the kidney tests, Urea is the dominant feature, as reflected in Fig. 1a where the box representing Urea is much larger than other boxes in this category. The boxes for Lactate and Linfocitos in the blood test category have comparable sizes displayed indicating that they have a similar impact on the ranking. The tool's semantic zoom feature allows users to expand the attributes of two or more patients to compare their feature importance. Users can also compare the feature values of selected patients on the histogram.

The treemap offers insight into the local important feature for each patient, providing users with a more granular understanding of the local feature importance influencing the patient's ranking. In contrast, the histogram showcases the top five globally important features, giving users an understanding of the feature importance across the whole population. By comparing feature importance at both the local and global level, users can develop a better understanding of the ranking system and the features that contribute to it.

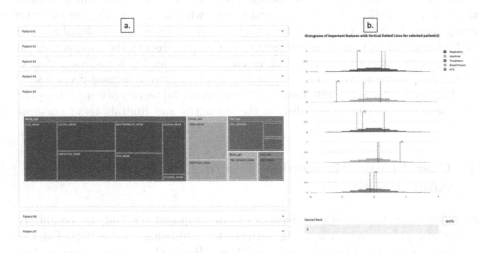

Fig. 1. An overview of the entire visual analytics tool including interactive ranking list and what-if panel, a) Interactive List of ranked patients and Treemap Visualization of Patient ranked 1 (P1) Features. b) What-if Panel with Histograms of 5 Top Important Features.

In the what-if panel depicted in Fig. 1b the blue dotted line denotes the selected patient (P5) feature value, while the black and red dotted lines indicate the amount of P5 feature that would result in a higher or lower ranking, respectively. The absence of a red or a black dotted line for each feature implies that modifying only this feature of the patient is insufficient to achieve a higher ranking, and modifying more than one feature (including this feature) is necessary. Our visualization tool allows users to set a desired ranking for a selected patient (in the text box) and presents the minimum changes required for the patient to attain that ranking, using the updated black and red lines. To illustrate, suppose a desired rank of 3 is entered for a patient currently ranked 5th. The updated red line in Fig. 1b represents the minimum changes required in features, namely Temperature, PCR, and HeartRate, to achieve a higher rank of 2 compared to the previous rank of 5. Conversely, modifying these features as indicated by the black lines results in a lower rank of 8. This example highlights how the visualization tool allows users to explore the impact of changing specific features on a patient's rank, providing valuable insights into the ranking system.

Overall, our proposed method effectively explained the ICU patient ranking system, enabling users to gain insights into how patients were ranked and explore potential changes that could impact the ranking.

5 Discussion

The use of conventional statistical methods such as accuracy, precision, and sensitivity is often not sufficient to provide users with a clear understanding of why a particular item has been ranked in a certain way. This is particularly true for ranking systems that use machine learning models, which can be difficult to interpret. An alternative approach that is applicable to a variety of ranking systems and empowers users to explore and understand the results of the ranking system is needed.

The proposed visual analytic tool is model agnostic and provides users with the ability to identify the contributing factors that determine the ranking of a specific item. By presenting the important feature contributors of prediction, users can understand what factors play a crucial role in determining the ranking of a particular item. The interactive visualization feature of the system allows users to gain a global understanding of how the system ranks items.

The Treemap helps users to understand what feature is important for an item to be ranked as it was. The expanding and collapsing attribute of the ranking list is another useful feature that allows users to drill down into more information as needed. This feature can be helpful when users need to see the distribution of important features and how a particular item ranks in comparison to others. Users can also see the changes needed to be ranked differently, providing them with actionable insights to improve the ranking of a specific item.

The what-if panel shows what features are essential for the whole population. This feature enables users to compare the ranking of a specific item to the overall ranking system and gain insights into the underlying factors that determine the ranking. Additionally, users can compare two or more items, enabling them to identify similarities and differences between them and understand the changes required for them to be ranked differently. Furthermore, users can investigate what changes are needed for an item to be ranked differently.

Overall, the proposed system provides users with a more transparent and interactive approach to ranking items. By providing users with a better understanding of the underlying factors that determine the ranking, the system can help users make more informed decisions and improve the overall quality of the ranking system. Therefore, the proposed system has significant potential to improve the usability and effectiveness of ranking systems in various fields.

6 Conclusion

In this paper, we proposed a visual analytic tool that combines XAI methods and interactive visualization to explain ranking systems by enabling users to investigate how changing the feature values of an item can impact the ranking. Our proposed method was evaluated through a case study on ICU patient triage. The case study demonstrated how our proposed tool can provide users with a better understanding of how ranking systems work, which can ultimately improve decision-making processes. The counterfactual explanation method allowed users to explore how changes to individual patient features could have resulted in a different ranking, while feature importance provided insights into the importance of different features in the ranking system. Additionally, the interactive visualization allowed users to easily explore and experiment with different scenarios. Overall, our proposed tool has the potential to be applied in various domains, such as healthcare, finance, and education, to improve transparency and trust in ranking systems. Future work will investigate the integration of XAI methods, such as fairness metrics and algorithmic auditing, with interactive visualizations to detect and mitigate bias in the ranking system.

References

1. Sivapalan, S., Sadeghian, A., Rahnama, H., Madni, A.: Recommender systems in e-commerce. In: 2014 World Automation Congress (WAC), pp. 179–184 (2014)
2. Rappaz, J. Dynamic personalized ranking. EPFL (2022)
3. Faliagka, E., et al.: On-line consistent ranking on e-recruitment: seeking the truth behind a well-formed CV. Artif. Intell. Rev. **42**, 515–528 (2014)
4. Yu, P., Lam, K., Lo, S.: Factor analysis for ranked data with application to a job selection attitude survey. J. R. Stat. Soc. A. Stat. Soc. **168**, 583–597 (2005)
5. Schoonderwoerd, T., Jorritsma, W., Neerincx, M., Van Den Bosch, K.: Human-centered XAI: developing design patterns for explanations of clinical decision support systems. Int. J. Hum Comput Stud. **154**, 102684 (2021)
6. Adadi, A., Berrada, M.: Peeking inside the black-box: a survey on explainable artificial intelligence (XAI). IEEE Access. **6**, 52138–52160 (2018)
7. Weng, D., Chen, R., Deng, Z., Wu, F., Chen, J., Wu, Y.: SRVis: towards better spatial integration in ranking visualization. IEEE Trans. Vis. Comput. Graph. **25**, 459–469 (2018)
8. Pereira, M., Paulovich, F.: RankViz: a visualization framework to assist interpretation of Learning to Rank algorithms. Comput. Graph. **93**, 25–38 (2020)
9. Gratzl, S., Lex, A., Gehlenborg, N., Pfister, H., Streit, M.: LineUp: visual analysis of multi-attribute rankings. IEEE Trans. Vis. Comput. Graph. **19**, 2277–2286 (2013)
10. Di Sciascio, C., Sabol, V., Veas, E.: Rank as you go: user-driven exploration of search results. In: Proceedings of the 21st International Conference on Intelligent User Interfaces, pp. 118–129 (2016)
11. Anahideh, H., Mohabbati-Kalejahi, N.: Local explanations of global rankings: insights for competitive rankings. IEEE Access **10**, 30676–30693 (2022)

12. Zhuang, H., et al.: Interpretable ranking with generalized additive models. Proceedings of the 14th ACM International Conference on Web Search and Data Mining, pp. 499–507 (2021)

13. Štrumbelj, E., Kononenko, I.: Explaining prediction models and individual predictions with feature contributions. Knowl. Inf. Syst. **41**, 647–665 (2014)

14. Karimi, A., Barthe, G., Balle, B., Valera, I.: Model-agnostic counterfactual explanations for consequential decisions. In: International Conference on Artificial Intelligence and Statistics, pp. 895–905 (2020)

15. Hashemi, M., Fathi, A. PermuteAttack: counterfactual explanation of machine learning credit scorecards. ArXiv Preprint ArXiv:2008.10138 (2020)

16. Mothilal, R., Sharma, A., Tan, C.: Explaining machine learning classifiers through diverse counterfactual explanations. In: Proceedings of the 2020 Conference on Fairness, Accountability, and Transparency, pp. 607–617 (2020)

17. Verma, S., Boonsanong, V., Hoang, M., Hines, K., Dickerson, J., Shah, C.: Counterfactual explanations and algorithmic recourses for machine learning: a review. arXiv preprint arXiv:2010.10596 (2020)

18. Maragno, D., Röber, T., Birbil, I.: Counterfactual explanations using optimization with constraint learning. ArXiv Preprint ArXiv:2209.10997 (2022)

19. Wachter, S., Mittelstadt, B., Russell, C.: Counterfactual explanations without opening the black box: automated decisions and the GDPR. Harv. JL Tech. **31**, 841 (2017)

20. Russell, C., Kusner, M., Loftus, J., Silva, R.: When worlds collide: integrating different counterfactual assumptions in fairness. In: Advances in Neural Information Processing Systems, vol. 30 (2017)

21. Dandl, S., Molnar, C., Binder, M., Bischl, B.: Multi-objective counterfactual explanations. In: Bäck, T., et al. (eds.) PPSN 2020, Part I. LNCS, vol. 12269, pp. 448–469. Springer, Cham (2020). https://doi.org/10.1007/978-3-030-58112-1_31

22. Yang, W., Li, J., Xiong, C., Hoi, S.: MACE: an efficient model-agnostic framework for counterfactual explanation. ArXiv Preprint ArXiv:2205.15540 (2022)

23. Dunsmuir, D.: Selective semantic zoom of a document collection, pp. 1–9 (2009)

24. Johnson, B.: TreeViz: treemap visualization of hierarchically structured information. Proceedings of the SIGCHI Conference on Human Factors in Computing Systems, pp. 369–370 (1992)

25. Sírio-Libanes Hospital. COVID-19 - Clinical Data to assess diagnosis. https://www.kaggle.com/datasets/S%C3%ADrio-Libanes/covid19. Accessed 20 Jan 2023

Explainable Artificial Intelligence in Response to the Failures of Musculoskeletal Disorder Rehabilitation

Laurent Cervoni[1], Rita Sleiman[1], Damien Jacob[1(✉)], and Mehdi Roudesli[2]

[1] Talan Research and Innovation Centre, Paris, France
{laurent.cervoni,rita.sleiman,damien.jacob}@talan.com
[2] Centre de l'appareil Locomoteur de l'Estuaire, Le Havre, France
m.roudesli@skeewai.com

Abstract. Osteoarticular pathologies, and particularly low back pain and ankle sprains, due to their number and recurrence, constitute a public health issue. Practitioners do not have enough data describing the impact of treatments on the evolution of pathologies, which is necessary to develop a program that can dynamically adapt to changing patient conditions. We have therefore designed an application based on a series of medical consensus rules capable of generating dynamic exercise sessions adapted to the patient's pathology and its evolution. In the traditional care pathway, it is difficult in retrospect to understand the failures in management (which amount to 40% in ankle sprains). Our approach, which adapts to the patient's state of health over time, allows us to better understand how exercises are generated and then to analyze the pathways in order to monitor their effectiveness.

The application, resulting from this work, is available as a WebApp.

Keywords: Musculoskeletal disorders · Artificial Intelligence · Health · Logical Programming · Consensus rules

1 Introduction

Osteoarticular pathologies of the lower limbs represent 25% of all osteoarticular pathologies. The ankle, with nearly 6,000 people affected [1] per day in France, alone represents 7 to 10% of hospital emergencies and more than a quarter of lower limb pathologies. It is a public health problem whose cost is estimated at 1.2 million euros per day. The total annual cost to society of ankle sprains has even been estimated at approximately 40 million euros per million people [1].

The number of recurrences speaks for itself: 30% of people have a persistent problem beyond 1 year [2, 3]. Other studies point to even more important after-effects. This underlines the importance of prevention and rapid treatment [4].

The responsibility of taking charge is generally based on "consensus rules" that allow the severity of the pathology to be assessed. Proprioception approaches, for example, have resulted in a 50% reduction in recurrences [5]. The role of self-education in the rehabilitation strategy is also emphasized.

J. M. Juarez et al. (Eds.): XAI-Healthcare/PM4H 2023, CCIS 2020, pp. 14–24, 2024.
https://doi.org/10.1007/978-3-031-54303-6_2

A comparison of supervised and self-rehabilitation tends to show similar functional results for the ankle. Better compliance was observed, particularly in terms of keeping medical appointments and better results in following the treatment program.

Low back pain is also a similar public health issue, with 80% of the population likely to suffer from low back pain in their lifetime, and more than half of the French population reporting an episode of back pain in a year [6]. As the first-line treatment, self-management and targeted exercises are recommended by the HAS (French National Authority for Health). For this pathology, several studies, implemented after Recov'Up (the application proposed in this paper, https://recovup.fr/), have aimed to demonstrate the interest of autonomous care, possibly assisted by an application [7, 8].

In this context, it would be interesting to be able to autonomize the patient in the realization of functional tests alone or assisted by computer, so that the practitioners have access to more information to personalize the care pathway.

Artificial intelligence can therefore play a significant role in supporting rehabilitation of osteoarticular pathologies and this empowerment strategy. However, even if AI has a growing number of medical applications, most of them are related to diagnostic or imaging assistance and much more rarely to therapy. One of the main reasons for this is the sensitive nature of medical data. Also, in many cases, there is no database available, which is the case for osteoarticular pathologies where practitioners rarely record the evolution of pathologies during rehabilitation sessions. This lack of data is one of the main motivations for imagining a system based on the practical rules of the medical profession, the effectiveness of these rules according to individual situations being then analyzed in a retrospective way.

Moreover, despite the advances in AI, the application of AI models still has limitations due to the fact that they are seen as black boxes, especially in domains such as medicine and healthcare, where explicability is essential to trust the results. Several studies are conducted to propose AI models with explainable capabilities in the healthcare domain [9, 10]. Mellem et al. [11] developed a methodology for choosing patients for clinical trials using the Bayesian Rule Lists algorithm to evaluate the effectiveness of schizophrenia treatment with a focus on model explicability. The authors found that applying their explainable rule-based AI approach to find more responsive patient subgroups increased treatment effects significantly and produced several logical statements that make interpreting the findings easier for clinicians. Gidde et al. [12] proposed CovBaseAI, an explainable COVID-19 detection system that uses both deep learning techniques and an expert rule-based system on chest X-Ray images, following radiologists' previously defined rules. The goal was to develop an AI solution capable of providing robust and explainable results with a smaller amount of new data.

In this paper, we present a practical case (Recov'Up) designed in the framework of a collaboration between the start-up Skeewai and the Research and Innovation Center of the Talan Group.

It is mainly based on consensus rules, the recommendations of the French National Authority for Health (Haute Autorité de Santé [13, 14]) and the expertise of practitioners (sports doctors and physiotherapists) expressed in the form of "logical" rules.

The proposed system (available as a WebApp) aims at a double level of understanding:

- Understand which rules have been implemented in the generation of an exercise sequence
- Identify the impact of the different sessions on the evolution of the pathology to improve the protocols.

2 Background and Context of This Work

In addition to the total cost to society, and to private companies, of treating pathologies such as ankle sprains and the resulting work stoppages, there is the risk of recurrence. Rapid re-mobilization of ankle sprains appears to reduce the risk of recurrence [15, 16].

However, only 50% of cases are subject to medical follow-up. A large proportion of patients are therefore at risk of developing chronic ankle instability [17, 18].

However, there is no universal "deterministic" medical method for treating ankle sprains. This situation is not surprising given the number of factors that come into play in the pathology and the very essence of medical practice, which must be as personalized as possible.

The Ameli website (French health insurance website) specifies, concerning rehabilitation, that "The recommended number of sessions is 10 for a recent ankle sprain. Rehabilitation is not systematic and is proposed according to the needs estimated by the doctor".

Thus, there are recommendations based on expert opinion, including the performance of tests to evaluate sensorimotor deficiencies. In particular, in the most recent work, which is the international reference for the analysis of ankle sprains, we observe that "The executive committee of the International Ankle Consortium has reached a consensus on recommendations for a structured clinical evaluation of acute lateral ankle sprain injuries. The recommendations are provided on the basis of an initial diagnostic clinical evaluation. The International Ankle Consortium ROAST [...] emphasizes the evaluation of mechanical and sensorimotor deficiencies known to be associated with chronic ankle instability". (Source Clinical assessment of acute lateral ankle sprain injuries (ROAST): 2019 consensus statement and recommendations of the International Ankle Consortium).

The absence of an "algorithmic" or simply formal approach leads to the consideration of experimental approaches that cannot be based on traditional computer programming.

It should be noted that despite the large number of cases, there is no database of patients with associated therapeutic follow-ups to date. An approach with a rule-based system therefore seemed legitimate. Indeed, in the medical field, Artificial Intelligence has been widely used for a long time, especially in imaging, pathology detection, or even diagnostic assistance. However, the exploitation of AI in the process of medical follow-ups is more rare.

The Recov'Up project was therefore initiated on the basis of these findings, at the beginning of 2020. The medical team collected all the exercises used by practitioners and available in the literature, regardless of whether or not simple equipment is used. The list of exercises used by the generator varies from twenty to more than fifty depending on the pathologies managed by the application.

The implementation of the exercises in the care pathway was defined by the medical team with regard to the benefit brought according to the initial date of the pathology, its nature, the patient's ability to perform certain simple functional tests and mainly the responses on the evolution of pain. The development of the engine modelling the exercises and the associated interface took about a year before the actual deployment.

3 Complexity of the Generation of Self-recovery Exercises

In the treatment of the ankle, we have about thirty exercises, classified in different families (based in particular on the recommendations of the High Authority of Health [19]). The objective is to propose sessions of 3, 4 or 5 exercises, chosen from this. The number of sessions is fixed by the rules proposed by the medical team according to the evolution of the pathology and avoiding monotony of the sessions in order not to tire the patient.

The complete space is theoretically:

$$C_{30}^3 + C_{30}^4 + C_{30}^5 = 4060 + 27\,405 + 142\,506 = 173\,971 \tag{1}$$

In practice, the real space is a little more complex to calculate, some exercises having, according to the situations, to be taken in a family or another. Some families are exclusive of the others or on the contrary can be associated.

If one introduces the notions of durations, the space is clearly more important. The management of duration means adding repetitions in order to fit into a time interval for a session to be acceptable to the patient with an effective minimum duration (a physiotherapist's session lasts between 15 and 20 min).

If we wish to simulate the management of exercises, in a discrete domain, this would mean adding to the list of exercises all the possible cases of repetitions. Since the exercises can be performed in sets of 8 to 15 repetitions, 3 to 5 times each, there would therefore be a space of 40 to 75 reps of each exercise. In this case, the total space exceeds several million possible combinations since all associations of repetitions would have to be calculated.

3.1 Defining the Rules

We have structured the generation of exercises into 4 categories of rules:

1) *General rules*:

- A session must contain between 3 and 5 exercises,
- The rules of progressiveness which specify the evolution of the number of exercises and series of an exercise must always be respected,
- A session must begin with a resting Visual Analog Scale (VAS) and end with a post-session VAS,
- Rehabilitation takes place over a minimum period of time (which may vary according to pathologies other than the ankle) and is only completed if a certain number of criteria are met.

2) *Rules of progressiveness*:

- The proposed exercises must come from at least 2 different families (these being, for example, proprioception, strengthening, eccentric, etc...),

- Before D5 (i.e. in the 5 days following the injury), apply the RICE protocol (Rest, Ice, Compression, Elevation),
- No exercise of families considered medically ineffective after D15,
- No overstressing exercise before D15,
- Never propose less than 2 exercises of certain families from D21,
- Etc…

3) *Irritability rules* such as: If the post-session VAS is higher than the resting VAS by more than 20% then do not follow the progressivity rule for the next session.

4) *"Cause and effect" rules*:

- Different families of exercises are proposed according to functional tests carried out and self-evaluated by the patient,
- The exercises are adapted based on the results of scores (such as the Limb Symmetry Index) taking into account whether or not the patient has the necessary equipment (elastic, unstable cushion).

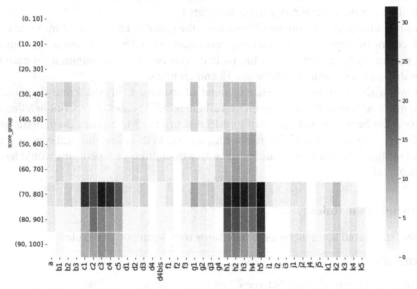

Fig. 1. The variation in FAAM (Foot Ankle Ability Measurement) scores, aggregated for different patient groups, across various exercise regimens for ankle sprain rehabilitation (FAAM scores on the vertical axis against different categories of exercise regimens on the horizontal axis).

3.2   Application of the Consensus Rules

Given the possible combinatorial nature of the exercises, it is important to understand what rules were applied. The application allows to record the proposed exercises with the associated scores (see Fig. 1).

By combining this explanation with the monitoring of the evolution of the pathologies, it becomes possible to determine the most efficient pathways according to the pathology and the characteristics of the patient.

4 General Structure of Recov'Up

The exercise generation engine corresponds to the modelling of the rules expressed by the doctors and physiotherapists involved in the project and to the search for a tuple satisfying these rules. The objective is to have an intelligent tool able to build sessions identical to those that a practitioner could propose by having a mechanism of explicability of the reasoning and by being more "imaginative" than a human by the aggregation of various categories of exercises.

Prolog was chosen for its ease of representing the descriptive constraints of the situation, its ability to infer on them to determine if a sequence of exercises could be proposed and its readability for the physicians. The program flow and the solutions can be explained, which makes it possible to add or change rules to be closer to the real situations of the patients.

Moreover, the application is perfectly compliant with the management of private data, as no personal data is required to generate sessions (in fact, gender, age or weight, for example, do not actually provide any medically useful information when the answers to the questions asked are available).

The patient's data is hosted on a Health Data Host and the Prolog engine is queried via a protocol that allows it to have the patient's history useful to suggest exercises but without needing any personal information, it then returns the answers to the query to the HDS.

This approach makes it possible to build a new database from which, in a second phase of the project, the algorithm can become even more relevant in its exercise suggestions.

The structure of the Prolog terms has been imagined in anticipation of an evolution towards other pathologies later on. The user interface is designed in PHP and the communication between PHP and Prolog is done via JSON.

The exercises are classified as standard Prolog facts and organized by family and their potential eligibility depends on the current day, the presence or not of specific equipment or the ability of the patient to perform certain physiological tests.

We thus find in Prolog, the rules as expressed by the health professionals who participated in the project (the Prolog code is in French):

```
exoeligible(Patient,cheville, J, i1, _) :-
      J>=15,
      exercicesCumules(Patient, cheville, _, ListeExo),
      comptemembre(h, ListeExo, Num),
      Num >=3,
      questiontest(Patient,cheville,J,coussin,_,o,_).

exoeligible(Patient,cheville, J, ExoFamilleK,_) :-
      J >= 21,
      questiontest(Patient,cheville,_,sport,_,o,_),
      questiontest(Patient,cheville,_,qf11,_,Repqf11,_),
      Repqf11 \= a,
      questiontest(Patient,cheville,_,qf12,_,Repqf12,_),
      Repqf12 \= a,
       questiontest(Patient,cheville,_,qf13,_,Repqf13,_),
      Repqf13 \= a,
      exercice(cheville, k, _, ExoFamilleK, _, _).
```

The exercise generator itself then exploits a "second-order logic capability" of Prolog by collecting all potentially eligible exercises with a bagof command:

```
bagof(Exo,   exoeligible(Patient,   Pathologie,   Jour,   Exo,_),
ListeExo)
```

The predicate "exoeligible" determines the applicability of an exercise and is thus triggered by "bagof" which fills the ExoList of all the Exos (French abbreviation for execice) corresponding to a given Patient for the current Day.

Fig. 2. Examples of exercises proposed for a patient with little pain and a favorable evolution of the pain.

The series of exercises considered as eligible is then filtered by additional rules (such as, for example, non-monotony to avoid a succession of identical exercises from

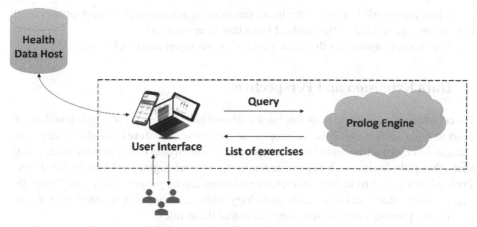

Fig. 3. General structure of Recov'Up.

one day to another) in order to generate a sequence that is transmitted via JSON to the interface module that displays the exercises of the day (see Fig. 2). The general structure of Recov'up is presented in Fig. 3.

In the extract above, we can see that the pathology is already anticipated as being able to have a value other than "cheville" (ankle), the system having been extended to 12 osteoarticular pathologies.

5 Explicability

The intrinsic readability of Prolog makes it possible to follow the "reasoning" when generating exercises. It is also possible to analyze the JSON to get the description of the patient's situation, the list of the proposed session, the total accumulation of exercises (in order to check the associated rules).

Although, in the chosen architecture, the information related to a patient is in a "secure enclave", at each request between the interface and the exercise generation, patient-specific information is transmitted to the Prolog engine. This makes it possible to dynamically reconstitute the patient's history and to have an explanation of the elements that led to the production of an exercise sequence or to the asking of specific functional questions:

```
?- evolution_eva(1736,cheville, EvolList).
EvolList = [68, [4, 5], 66, [3, 3], 64, [6, 5], 62, [24, 31]].
```

This gives the evolution of the VAS values before and after each exercise session for the indicated days (68, 66, etc...).

```
?- material(1736,cheville, 68, X).
X = cushion.
```

Allows you to make sure that patient 1736 had a cushion to do exercises on day 68.

It is also possible to retrieve the list of functional questions that the patient answered, the sessions generated or the medical rules that were retained.

The protocol applied is therefore perfectly transparent and can be justified.

6 Data Extension and Perspectives

Faced with pathologies such as low back pain and ankle sprains, for which few datasets exist to build accurate models, we adopted an approach consisting in implementing rules recognized by practitioners. The exercise generation engine is designed in such a way that other pathologies, such as patellar tendinitis or cervicalgia, can be easily integrated. Prolog has allowed us to have an incremental approach by progressively completing the engine with rules specific to each pathology while ensuring not to interfere with the others and guaranteeing the strict application of these rules.

The proposed application, Recov'Up, offers the possibility to explain its decisions, which facilitates its adoption by health professionals and makes it possible to face the lack of medical data for these pathologies. Moreover, the architecture chosen does not use personal information, as this does not provide any information for the treatment of pathologies: only the answers to the functional questions are useful to evaluate the patient's condition and to decide on the exercises to be performed. This approach facilitates the analysis of the data collected by various external complementary tools.

Moreover, Recov'Up allows the collection of anonymized data, which can be analyzed to improve the recommendations of exercise sequences. The collected data include patient information such as age, gender, height, weight, pain on day D, pain on day $D + 1$, pathology, and recommended exercises. The methodology aims to achieve the following objectives:

- Identifying Effective Exercises: Using the collected data to assess the effectiveness of exercises and to filter out the exercises that had no positive or negative effects on the healing process,
- Developing a ML-Based Recommendation System: Automatically recommending effective exercises without predefined rules.

The process begins with the application of an unsupervised clustering algorithm, like K-means, to group patients based on age, gender, height, weight, pain on day D, and pathology, resulting in K clusters. This helps us identify K prototypes of patients: the centers of the K clusters. New patients are assigned to the appropriate cluster, and their recommended exercises align with those of historical patients in the same cluster.

To eliminate ineffective exercises, the average pain evolution for each exercise is calculated for each prototype group, dividing pain on day $D + 1$ by pain on day D. Exercises with an evolution greater than or equal to one are considered ineffective and are removed from the list. Finally, additional filtering is applied to the eligible exercise list, incorporating rules such as avoiding monotony to prevent the repetition of identical exercises and considering the presence of equipment, as implemented in the current application version.

Another ML application crucial in our context involves predicting the recovery pattern of each new patient. Using K-means to cluster patients based on their recovery pattern (pain evolution), we can discern the most significant recovery patterns. Subsequently, a ML classification algorithm, such as Random Forest, is trained to predict the recovery pattern for each new patient based on their individual information.

The selection of ML algorithms, namely K-means and Random Forest, is significant for considering the interpretability of the final proposed system. K-means is known for its transparency, characterized by a straightforward inner algorithm. On the other hand, while each tree in the Random Forest is individually interpretable, the combination of all trees yields a more complex model. However, by leveraging techniques such as features importance and post-hoc explainable AI methods like SHAP values, we can gain insights into the contribution of each input feature to the final decision.

The main obstacle preventing the implementation of these proposed systems is the challenge of data collection. Training these systems effectively relies on a substantial amount of reliable data. However, a significant issue arises as some patients initiate the process but fail to adhere consistently, either by not completing all recommended exercises or discontinuing their participation altogether. This lack of uniformity in patient engagement poses a considerable challenge in assembling a comprehensive and reliable dataset necessary for robust algorithmic training. Overcoming this limitation is crucial to ensuring the algorithm's accuracy, robustness and applicability.

In an effort to measure the effectiveness of the proposed model, we are conducting an additional study aimed at developing a computer vision model. This model is designed to monitor and measure patient movements, which will help their recovery progress and the accuracy of their exercise execution. It is expected that this tool will markedly advance our ability to evaluate the success of our proposed solution.

Disclosure of Interests.   The authors have no competing interests to declare that are relevant to the content of this article.

References

1. Avis de la CNEDiMTS (2018). https://www.has-sante.fr/upload/docs/evamed/CEPP-5487_A2T_15_mai_2018_(5487)_avis.pdf
2. Van Rijn, R.M., Van Os, A.G., Bernsen, R.M., Luijsterburg, P.A., Koes, B.W., Bierma-Zeinstra, S.M.: What is the clinical course of acute ankle sprains? A systematic literature review. Am. J. Med. **121**(11), 324–331 (2008)
3. Polzer, H., et al.: Diagnosis and treatment of acute ankle injuries: development of an evidence-based algorithm. Orthop. Rev. **4** (2012)
4. Wagemans, J., Bleakley, C., Taeymans, J., et al.: Rehabilitation strategies for lateral ankle sprain do not reflect established mechanisms of re-injury: a systematic review. Phys. Ther. Sport (2023)
5. Hupperets, M.D., Verhagen, E.A., Van Mechelen, W.: Effect of unsupervised home based proprioceptive training on recurrences of ankle sprain: randomised controlled trial. Bmj **339** (2009)
6. La lombalgie, un enjeu de santé publique. https://www.ameli.fr/medecin/sante-prevention/pathologies/lombalgies/enjeu-sante-publique
7. Fatoye, F., Gebrye T., Fatoye C., et al.: The clinical and cost-effectiveness of telerehabilitation for people with nonspecific chronic low back pain: randomized controlled trial. J. Med. Internet Res. mHealth uHealth **8** (2020)

8. Itoh, N., Mishima, H., Yoshida, Y., et al.: Evaluation of the effect of patient education and strengthening exercise therapy using a mobile messaging app on work productivity in Japanese patients with chronic low back pain: open-label, randomized, parallel-group trial. J. Med. Internet Res. mHealth uHealth **10** (2022)
9. Tjoa, E., Guan, C.: A survey on explainable artificial intelligence (XAI): toward medical XAI. IEEE Trans. Neural Netw. Learn. Syst. **32**(11), 4793–4813 (2021)
10. Loh, H.W., Ooi, C.P., Seoni, S., Barua, P.D., Molinari, F., Acharya, U.R.: Application of explainable artificial intelligence for healthcare: a systematic review of the last decade (2011–2022). Comput. Methods Program. Biomed. (2022)
11. Mellem, M.S., Kollada, M., Tiller, J., Lauritzen, T.: Explainable AI enables clinical trial patient selection to retrospectively improve treatment effects in schizophrenia. BMC Med. Inf. Decis. Making **21** (2021)
12. Gidde, P.S., et al.: Validation of expert system enhanced deep learning algorithm for automated screening for COVID-Pneumonia on chest X-rays. Sci. Rep. **11** (2021)
13. HAS: Prise en charge du patient présentant une lombalgie commune (2019). https://www.has-sante.fr/upload/docs/application/pdf/2019-04/fm_lombalgie_v2_2.pdf
14. Rééducation de l'entorse externe de la cheville (2000). https://www.has-sante.fr/upload/docs/application/pdf/recosentors.pdf
15. Chen, E.T., Mcinnis, K.C., Borg-Stein, J.: Ankle sprains: evaluation, rehabilitation, and prevention. Curr. Sports Med. Rep. **18**(6), 217–223 (2019)
16. Hoch, M.C., Hertel, J., Gribble, P.A., et al.: Effects of foot intensive rehabilitation (FIRE) on clinical outcomes for patients with chronic ankle instability: a randomized controlled trial protocol. BMC Sports Sci. Med. Rehabil. **15**(1), 1–13 (2023)
17. Pijnenburg, A.C., Van Dijk, C.N., Bossuyt, P.M., Marti, R.K.: Treatment of ruptures of the lateral ankle ligaments: a meta-analysis. J. Bone Joint Surg. **82**(6), 761–73 (2000)
18. Gribble, P.A., et al.: 2016 consensus statement of the international ankle consortium: prevalence, impact and long-term consequences of lateral ankle sprains. Br. J. Sports Med. **50**, 1493–1495 (2016)
19. 2004 update of the consensus conference ankle sprains in the emergency department. https://www.sfmu.org/upload/consensus/actualisation_entorse.pdf

An Explainable AI Framework
for Treatment Failure Model for Oncology
Patients

Syed Hamail Hussain Zaidi(✉) ⓘ, Bilal Hashmat, and Muddassar Farooq

CureMD Research, 80 Pine St 21st Floor, New York, NY 10005, USA
hamail1041@gmail.com, {bill.hashmat,muddassar.farooq}@curemd.com
https://www.curemd.com/

Abstract. The black box nature of current AI models has raised serious
concerns about accountability, bias and trust in the models that might
undermine their relevance and usefulness in the field of medicine where
human lives are at risk. AI in medicine has the ability to derive meaningful inferences from real world data – an emerging school of thought
namely Real World Evidence (RWE) studies – that can assist medical practitioners to improve evidence based quality of care. In the field
of oncology, the accuracy and performance of inference models are as
important as clinically relevant and sound explanations of the inference.
In this paper, we present an Explainable AI (XAI) framework for our AI
model that predicts the suitability of a chemotherapy treatment at the
time of its prescription based on RWE. The framework provides explanations both for a specific patient and also for the model. It provides
explanations like feature analysis, counterfactual, and top risk factors
that contribute to a treatment failure. As a result, the framework adds
an explainability layer between treatment failure predictive model and
oncologists, thereby enabling evidence based assistance to oncologists in
designing chemotherapy plans.

Keywords: Explainability · Machine Learning · Model Agnostic ·
Oncology

1 Introduction

The black box nature of current AI models has raised serious concerns about
accountability, bias and trust in the models that might undermine relevance
and usefulness in the field of medicine where human lives are at risk. AI in
medicine has the ability to derive meaningful inferences from real world data –
an emerging school of thought namely Real World Evidence (RWE) studies –
that can assist medical practitioners to improve evidence based quality of care. If
these models make incorrect or biased predictions, a patient's health and safety

Supported by CureMD.

is compromised. Therefore, it is important to ensure that healthcare providers are empowered to understand the reasons behind an inference or prediction and whether the reasons are clinically sound or not. As a result, they can feel more confident and put trust in the accuracy and trustworthiness, hence ensuring positive outcomes for patients' health and safety.

Using AI in Oncology for predicting treatment failure of chemotherapy plans is a challenging problem, as more than 10% patients die because of side effects of chemotherapy [1] and not from the cancer itself. In an earlier work, not in the scope of this paper, we have developed a treatment failure AI model that predicts with more than 80% accuracy whether a chemotherapy treatment plan would discontinue or fail during its planned duration. However, oncologists in our partner provider network, are emphasizing the need that the AI model must also explain the reasons behind its prediction, so that they feel comfortable using the system if it generates clinically sound and relevant explanations. As a result, oncologists and the AI assistant – named MedicalMind – would work together as a team to select chemotherapy plans that have a low probability of failure, which will significantly reduce the financial and emotional toxicity of treatment failure on patients, their families and the healthcare system. Oncologists can always overrule a recommendation, if they think the reasons are not clinically sound or relevant making MedicalMind a truly Human Centered AI System that avoids causing harm to cancer patients.

The major contributions of this paper are:

- A model agnostic explainability module that provides meaningful explanations about treatment failure to oncologists about the inner workings of the model;
- A patient specific explainability module that generates meaningful explanations based on a patient's profile that contribute towards treatment failure by showing top risk factors and features that contribute towards treatment failure;

2   Scope of Work

In this paper, we adopted the explainability framework presented in [2] as a template for our treatment failure module [3] in the oncology domain and it is built on three pillars.

Algorithm Transparency. The understanding of what relationships a machine learning model has learnt and how the model uses them to make treatment failure predictions. It provides a general understanding of inner workings of a machine learning model.

Global Interpretability. The explanation of how the treatment failure model makes a failure prediction by analysing the distribution of the target outcome based on features. It depends on data and RWE, and the machine learning model. This shows important features that play a role in predictions.

Local Interpretability. It explains why a model predicted a certain outcome based on electronic health record of a patient. For a patient, certain features may have a big influence on a prediction. In medicine, local interpretations that are relevant to a patient are more desirable and considered important than global explanations.

2.1 Approaches to Explainability

In [2], the authors suggest to use those algorithms that generate interpretable models like linear regression, logistic regression and simple decision trees. One limitation of using such algorithms is their inability to learn complex decision boundaries. Model-agnostic methods of interpretability help in overcoming this challenge, as they are not tightly bound to a model. Separating the model from the explanations helps in creating a flexible XAI framework where developers can generate explanations for any machine learning model.

Shap. The basic idea behind SHAP, presented in [4], is to use game theory and the concept of Shapley values to assign values to each input feature, representing its contribution to the final prediction. The Shapley value is a way of distributing the total gain or cost of a group effort among the participants based on their individual contributions. In the context of machine learning, the players can be considered as input features, and the gain is the model output. Theoretically, shapley value for a feature j can be defined as follows:

$$\phi_j = \frac{1}{M} \sum_{S \subseteq \mathcal{F} \setminus j} \frac{|S|!(|\mathcal{F}| - |S| - 1)!}{|\mathcal{F}|!} [f(S \cup j) - f(S)] \tag{1}$$

where $\mathcal{F}$ is the set of input features, M is all possible subsets of $\mathcal{F}$, f is a mapper function that maps subset of features to model output, and $|S|$ is the number of features in one subset S. The term $[f(S \cup j) - f(S)]$ represents the change in the model's output when feature j is added to the subset S. Intuitively, the shapley value for any feature j can be imagined as its average marginal contribution across all possible coalitions of features. It measures how much a prediction decision changes on an average when feature j is included, taking into account all possible combinations of other features.

Counterfactuals. Counterfactual reasoning are based on: "If X had not occurred, Y would not have occurred". This technique was first introduced in [5]. Counterfactual aims to explain why a particular prediction was made by providing a perturbed input that has changed the prediction class. It can help a user understand a minimal set of changes to the input feature space that would change the prediction from the original class to a desired class. Theoretically it can be explained as: let f be the complex model to be explained, and let x be the input instance that produced a prediction $y = f(x)$. A counterfactual explanation aims to find an input x' that is as close as possible to x, but produces a different prediction $y' = f(x')$. The minimal set of changes can be found by minimizing a distance metric between x and x' subject to the condition that $f(x') \neq y$ and that the changes are minimal.

Permutation Feature Importance. This method is used to compute the importance of features in a machine learning model. It works by randomly toggling the values of a feature and observing the effect on the model's performance. The decrease in performance after the change shows the importance of a feature in a model. Greater the drop in performance, the more important the feature is for a predictive model [6].

3 Methodology

The medical records of cancer patients, containing medical profiles and treatments administered, are provided by CureMD, after anonymizing and deidentifying them as per HIPAA guidelines for this research. We analysed the electronic health records of 5 most prevalent cancer types – Colon, Lung, Breast, Prostate, and Multiple Myeloma – to discover hidden trends, anomalies, and other insights to build RWE. Then, SHAP and counterfactual are implemented to generate patient specific explanations and permutation feature importance is used for global explanations. All explanations are integrated in our Explainable AI (XAI) framework as illustrated in Fig. 1. Our machine learning model makes the prediction about treatment failure using patients' history and chemotherapy plans, both prescribed and administered by oncologists, and this prediction is then fed into our XAI framework to generate explanations and present them in a user friendly fashion to oncologists, who can verify clinical soundness and relevance.

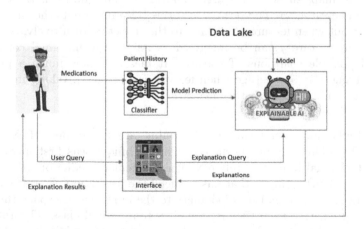

Fig. 1. Block Diagram of Oncology XAI Framework.

Data Analysis. For each cancer type, we created 3 splits of data: training, validation and test sets. The datasets of each cancer type are analyzed to detect any abnormalities, hidden trends, and bias in the data. Dataset insights are very important to make sure that the model has learnt the sound trends that are not biased to any specific feature. Moreover, treatment failure model did not include social and ethnic factors as features to remove the bias in the model. During this phase, distribution of the each feature of a cancer type is observed to check for outliers or skewness in data. To fix the outliers, multivariate imputer model is used to do imputation which estimates the values using an entire set of available feature dimensions and not just the mean value.

3.1   Treatment Failure Explanations

To generate treatment plan specific explanations, two model agnostic approaches are used: Shap and Counterfactual. The details are provided in the following.

Shapely Values. To find a suitable Shap explanation algorithm, different explanation algorithms such as TreeShap, KernelShap, DeepShap etc. are analyzed and KernelShap is then selected to generate Shapely values. DeepShap is optimized for deep learning algorithms and TreeShap is specifically for tree based models. Although, we are using boosted random forest model for predicting treatment failure which is a tree based model but TreeShap in Python does not yet have an implementation for a Random Forest. We have generated three different types of explanations from Shapely values:

- Feature importance in percentages;
- Features responsible for an increase in treatment failure probability;
- Features responsible for a decrease in treatment failure probability.

Counterfactual. To generate counterfactual, DiCE (Diverse Counterfactual Explanation) module of python is utilized. The module also has different algorithms to find counterfactual features. *Random* sampling algorithm is used as it is a simple and efficient method for finding counterfactual features. In this method, the algorithm randomly samples inputs from the search space and checks if they satisfy a given set of constraints. The search space is defined as the space of feature values that are close to the original values, but after satisfying constraints they produce a different prediction.

For perturbation of feature values, it is important to distinguish between categorical and continuous features. Continuous features are relevant here, as a feature that has more than 10 unique values is assumed to be continuous. Only chemotherapy treatment plan relevant features are picked for generating a counterfactual that reduces the treatment failure probability and this is something that an oncologist might alter at the time of designing a chemotherapy plan.

3.2   Model Level Explanations

Permutation feature importance and global Shapely values are being used to explain the inner workings of our treatment failure model.

Permutation feature importance scores are calculated and the global shapely values are the mean of shapely values of the complete dataset. The most important and the least important features are identified from these two XAI approaches. These attributions are then converted into percentage attribution and a threshold for choosing least important features of 0.2% is chosen. This threshold is not fixed and can be changed. Since both techniques use different methods to identify feature importance; therefore, combining them in a team to decide most and least important features is a significantly better option.

Global explanations do not need to be generated at run time because they will not change unless we change the treatment failure model or the oncology datasets. Therefore, global feature importance scores are computed and stored in a database. They will be recomputed and stored in a database if the model and dataset are updated.

3.3   Challenges and Limitations

We faced a number of challenges once we implemented our XAI framework. For example, KernelExplainer that generates shapely values is relatively slower compared with TreeShap and other algorithms because they are model specific optimized algorithms but do not support the boosted random forest model. For counterfactual, *generic* sampling returns the most optimized counterfactual but for some patients it takes a significantly large time because it keeps trying to find a better counterfactual than the one it has already found. It takes somewhere from one minute to 10 min to generate counterfactual with *generic* sampling. To cater for this shortcoming, we used *random* sampling method. This sampling algorithm is, sometimes, unable to find any counterfactual in the first attempt. Therefore, we try 5 to 6 times before raising an exception and even then this takes less time than *generic* method.

4   Results and Discussion

4.1   Dataset Insights

Insights derived from analyzing oncology datasets are discussed for only three cancer types due brevity. The prevalence of failed (or discontinued) chemotherapy plans in Colon, Lung, and Breast cancer are shown in Fig. 2.

Analyzing gender prevalence is also important to avoid any gender bias in the model. Analysis of Fig. 3 shows that no bias exists in the gender feature of a model. C50 is Breast cancer and its prevalence is 97% for female patients and this is supported by clinical studies in earlier work.

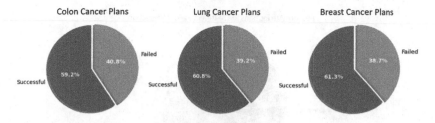

Fig. 2. Chemotherapy Plan Failure ratio in Oncology datasets

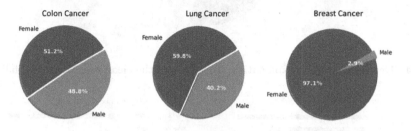

Fig. 3. Gender Prevalence in data sets

4.2 Treatment Failure Explanations

We now present a use case of a patient to discuss explanations of treatment failure for illustration only; nevertheless, the XAI framework is able to generate these types of explanation for all patients of all 5 cancer types. For example, a patient aged 62, diagnosed with stage 2 colon cancer is picked at random for our study. Treatment failure probability, predicted by our model, for a chemotherapy plan that is selected by an oncologist is 79.42%. Three different explanations are generated from the shapely values of patient's demographics: (1) percentage feature attributions, (2) top risk factors, and (3) safe features.

Percentage Feature Attributions. This feature analysis gives an insight to the important factors that led to a treatment failure. In our case, the top ten most important features that lead to predicting treatment failure for the patient mentioned in the above are shown in Fig. 4. For this patient, tumour size, overall stage and an earlier administered drug (Corticosteroids) are the top 3 features for the predictive model. Top 3 features may not necessarily contribute in failure prediction only.

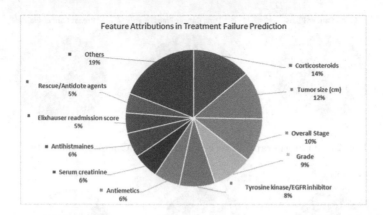

Fig. 4. Treatment specific explanation: Top features that lead to a treatment failure

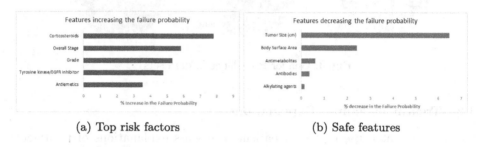

(a) Top risk factors   (b) Safe features

Fig. 5. Top features impacting treatment failure probability

Top Risk Factors. Oncologists are also interested in knowing the specific features in patient's history, demographics, diagnoses, or drugs that significantly contribute towards treatment failure. Risk factors calculated from the shapely values provide these explanations. Features that are causing % increase in the failure probability are shown in Fig. 5a. Interestingly, Corticosteroids and overall stage are also top 2 risk factors as well.

Safe Features. These features contribute towards lowering the failure probability. For our test case, the features causing % decrease in the failure probability are shown in Fig. 5b. It is interesting that the tumour size is not a risk factor.

4.3   Counterfactual

Our counterfactual module generates a diverse set of counterfactual features for a chemotherapy plan for a patient if its original failure probability is greater than 50%. Counterfactuals are generated for this patient and are tabulated in Table 1. Using the counterfactual, the recommended dosage of chemotherapy drugs suggested to be changed and with these new doses, the treatment failure probability is reduced to *43.53%*. Consequently, the counterfactual reduced the probability of treatment failure by *35.89%*. This feature helps an oncologist to adjust the dosages in a chemotherapy plan so that the probability of treatment failure is significantly reduced.

Table 1. Generated Counterfactual that reduce treatment failure probability

Drugs	Planned Dosage (g)	Recommended Dosage (g)
Antimetabolites	4.72	4
Angiogenesis inhibitors	0.27	1.07
Antibodies	0	2.1
Topoisomerase I inhibitors	0.3	0
Rescue/Antidote agents	0.67	1.07
Corticosteroids	0	0.3
Antiemetics	0.02	0

4.4   Model Level Explanations

Feature analysis and top risk factors that contribute in treatment failure are identified from permutation feature importance and global shapely values. The results of model explanations for each of the 5 studied cancer types are presented in Table 2.

Table 2. Model Level Explanations of treatment failure in different Cancer types

Cancer Type	Contributing Factors in Prediction		Top Risk Factors	
	Feature	Importance (%)	Feature	Risk Weight (%)
Colon	Antimetabolites	9.8	Antiemetics	21.21
	Body surface area	6.8	Corticosteroids	15.25
	Age	6.6	Angiogenesis inhibitors	13.45
	Rescue/Antidote agents	5.4	Rescue/Antidote agents	11.90
	Angiogenesis inhibitors	4.8	Overall Stage	9.38
	Elixhauser readmission score	4.6	Antiemetics	9.02
	Change in Elixhauser score	4.5	Elixhauser readmission score	6.43
	Antiemetics	4.3	Topoisomerase I inhibitors	5.06
	Corticosteroids	4.3	Tyrosine kinase/EGFR inhibitor	5.06
	Alkylating agents	4.1	Serum creatinine	3.23
Lung	Serum creatinine	6.14	Overall Stage	23.62
	Tumor size (cm)	6.13	Corticosteroids	15.29
	Antiemetics	5.2	Antimetabolites	12.21
	Alkylating agents	4.34	Age	11.24
	Overall Stage	4.26	T stage	10.40
	Corticosteroids	4.2	Elixhauser readmission score	9.34
	Body surface area	4.1	Small cell	5.07
	Age	3.93	Angiogenesis inhibitors	4.60
	Corticosteroids	3.88	Antiemetics	4.47
	Mitotic inhibitors	3.88	Anemia identified	3.72

(*continued*)

Table 2. (*continued*)

Cancer Type	Contributing Factors in Prediction		Top Risk Factors	
	Feature	Importance (%)	Feature	Risk Weight (%)
Breast	Endocrine and metabolic	9.61	T stage	21.6
	Serum creatinine	5.9	Metastasis	18.36
	Radiation therapy	5.69	Serum creatinine	16.4
	T stage	5.09	Overall Stage	10.9
	Age	4.89	Endocrine and metabolic	7.5
	Body Surface Area	4.69	Antiemetics	6.9
	Antiemetics	4.36	Elixhauser readmission score	5.3
	Antihistmaines	4.13	Antibiotics	4.7
	Elixhauser readmission score	3.96	Tumor size (cm)	4.2
	Alkylating agents	3.94	Antimetabolites	4.2
Prostate	LHRH analogs	16.6	Antiemetics	31.9
	Metastasis	9.6	LHRH analogs	14.03
	Age	7.1	Metastasis	11.10
	Body surface area	6.2	Antiemetics	9.78
	Tumor size (cm)	5.99	Elixhauser readmission score	8.80
	Serum creatinine	5.4	Mitotic inhibitors	7.27
	Antiemetics	3.3	T stage	6.91
	Elixhauser readmission score	3.1	Age	3.80
	Mitotic inhibitors	2.98	Alkylating agents	3.53
	Corticosteroids	2.90	Grade	2.86
Multiple Myeloma	Age	7.25	Enzyme inhibitors	17.01
	Endocrine and metabolic	7.18	Histone deacetylase inhibitors	13.5
	Histone deacetylase inhibitors	7.1	Antihistmaines	12.67
	Radiation therapy	5.08	Overall Stage	12.31
	Change in Elixhauser score	5.05	Endocrine and metabolic	12.29
	Body surface area	4.7	Corticosteroids	7.88
	Serum creatinine	4.45	Serum creatinine	7.18
	Enzyme inhibitors	4.32	Age	7.04
	Antihistmaines	4.2	Antibodies	5.79
	Elixhauser readmission score	3.35	Mineral/Electrolytes	4.32

5   Conclusion and Future Work

Our explainability framework is able to provide global and local explanations for a black-box type treatment failure AI model. A Model agnostic XAI is developed that adds an explainability layer between an AI treatment failure model and an oncologist to win their trust by providing clinically sound and relevant explanations. Moreover, the framework also provides top features and top risk factors that might lead to treatment failure for a given patient; as a result, an oncologists using counterfactual could review changes in dosages of drugs (or add/delete drugs) so that the probability of treatment failure is reduced. Consequently, the AI treatment failure model, along with proposed XAI framework, works as a clinically relevant RWE based chemotherapy assistant in planning the chemotherapy for patients for which a decision is not straightforward. The future work aims at generalizing the framework to other types of chronic diseases like diabetes and hypertension etc.

References

1. O'Brien, M.E.R., et al.: Mortality within 30 days of chemotherapy: a clinical governance benchmarking issue for oncology patients. Br. J. Cancer **95**(12), 1632–1636 (2006)
2. Mohseni, S., Zarei, N., Ragan, E.D.: A multidisciplinary survey and framework for design and evaluation of explainable AI systems. ACM Trans. Interact. Intell. Syst. (TiiS) **11**(3–4), 1–45 (2021)
3. Shahid, M.U., Farooq, M.: Boosted random forests for predicting treatment failure of chemotherapy regimens. In: Juarez, J.M., Marcos, M., Stiglic, G., Tucker, A. (eds.) AIME 2023. LNCS, vol. 13897, pp. 13–24. Springer, Cham (2023). https://doi.org/10.1007/978-3-031-34344-5_2
4. Lundberg, S.M., Lee, S.-I.: A unified approach to interpreting model predictions. In: Advances in Neural Information Processing Systems, vol. 30 (2017)
5. Mothilal, R.K., Sharma, A., Tan, C.: Explaining machine learning classifiers through diverse counterfactual explanations. In: Proceedings of the 2020 Conference on Fairness, Accountability, and Transparency, pp. 607–617 (2020)
6. Altmann, A., Toloşi, L., Sander, O., Lengauer, T.: Permutation importance: a corrected feature importance measure. Bioinformatics **26**(10), 1340–1347 (2010)

Feature Selection in Bipolar Disorder Episode Classification Using Cost-Constrained Methods

Olga Kaminska[1]([✉])[iD], Tomasz Klonecki[2][iD],
and Katarzyna Kaczmarek-Majer[1][iD]

[1] Systems Research Institute Polish Academy of Sciences, Warsaw, Poland
{o.kaminska,k.kaczmarek}@ibspan.waw.pl
[2] Institute of Computer Science, Polish Academy of Sciences, Warsaw, Poland
tomasz.klonecki@ipipan.waw.pl

Abstract. An important step in the classification process of bipolar disorder episodes is feature selection process indicating the most relevant factors in patients' behavior. The features in this task are associated with costs. Besides basic (low-cost) information about patients' phone calls and text messages, we are studying the impact of acoustic features (high-cost) on classifying patients' states. Unlike in previous papers, now we take the costs into account and thus we apply cost-constrained methods. The purpose of this paper is to examine whether the cost-constrained feature selection procedure is capable of improving the performance of the classification model while reducing the cost of making predictions. Moreover, we are trying to determine whether the reduced number of expensive features maintains a relatively high performance. We use a filter feature selection method that applies information theory. In the cost-constrained modification, we add a cost factor parameter that controls the trade-off between feature importance and its cost. The experiments were performed on a large medical database collected from patients with bipolar disorder during their daily mobile calls. The results indicate that the cost-constrained method allows to achieve better results than traditional feature selection when the budget is limited.

Keywords: bipolar disorder episode classification · mobile calls acoustic features · feature selection · information theory · costly features

1 Introduction

In Bipolar Disorder (BD) there are plenty of factors that have a significant impact on current patients' mood. The habit of using mobile phones is one of them. Both the behavioral features related to the frequency of using a mobile phone and the acoustic parameters from the patient's recordings seem to be reasonable markers that can help determine the current state of the patient. The

J. M. Juarez et al. (Eds.): XAI-Healthcare/PM4H 2023, CCIS 2020, pp. 36–40, 2024.
https://doi.org/10.1007/978-3-031-54303-6_4

selection of the most relevant features is a fundamental stage of data modeling and should be given a lot of attention from the very beginning. In medical problems, acquiring important features is associated with a financial cost. Current research aims to analyze whether the acoustic features (with the highest acquisition cost) are all important and demanding to maintain the high performance of classifying the BD state. Moreover, we want to indicate the most significant features that are capable of distinguishing the phase of BD patients. This research is an extension of our latest paper presented during the AIME'19 workshops [8], where the process of feature selection was conducted using manual analysis. It is an important step in the classification process of BD episodes, therefore we apply a cost-constrained automatic feature selection (FS) method, a model-independent filter algorithm. Its objective is to select a subset of features that are the most relevant and at the same time, their summarized cost does not exceed the user-defined budget. Each feature is related to a specific cost that is comparable to the financial cost of acquiring its value for a single observation. In this type of algorithm, it is critical to find the trade-off between the relevancy of the feature subset and its cost.

Some related research indicates that behavioral data collected by mobile phones [4] as well as acoustic features [9] based on mobile recordings are essential factors in the classification of BD episodes. Both authors indicate important features for particular BD states, regardless of the cost of acquiring them. The field of cost-constrained feature selection has primarily focused on the adaptation of traditional algorithms to be cost-constrained. There are modifications to the various filter methods [1,7,11]. In this work, we focus on the filter methods based on the information theory, essentially due to their model-independent nature and ability to detect non-linear interactions between features. For a general review of FS methods, we refer to [3] and in a particular case of a medical project that collects a large amount of data, we refer to [10].

2   Methodology

2.1   Data Preprocessing

The used BDMON [8] dataset received with OpenSmile [5] library describes patients diagnosed with BD based on their daily mobile phone calls. The final dataset contains 86 data streams with the main acoustic characteristics of the voice, such as loudness, voice energy or pitch and 11 behavioral features such as the number of incoming/outgoing calls per day or the average length of characters in daily text messages. That set of features was next smoothed in the time domain by applying an average of the day's measurements. Labels that were collected during patients' visits represent 4 bipolar phases: Depression, Mania, Euthymia and Mixed. They were extended for 7 days before the visit and 2 days after the visit [6].

We assume that two types of features: behavioral and acoustic have two different costs assigned. The standard application library[1] was able to collect

---

[1] Android Developer: https://developer.android.com/.

basic phone call statistics at the time the research started (2017) therefore, we assume that the cost of this feature type is equal to 1. For commercial use, the OpenSmile library is an expensive tool. We initially set the cost of each acoustic feature to 5 concerning the time that experienced programmers would need to extract them from mobile recordings. The assigned cost values could be controversial, thus we decided to compare two variants of them (5 and 10). Then compare the results based on the AUC score using Wilcoxon signed rank test, if there is a relevant difference between contentious costs values. We received p value > 0.05, so it indicates that there is no significant difference between the two distributions of firstly selected features therefore, we decided to use the cost equal 5 as a strategy in this research.

2.2 Algorithm

In this paper, we focus on the greedy forward selection algorithm based on information theory [11]. Let $X_1, \ldots, X_p$ be the features, $F = \{1, \ldots, p\}$ be the set of features indexes, Y be a multiclass variable that we consider as a target and $c(k)$ be the cost of the k -th feature. In each step of the algorithm, we add the index of the candidate feature X_k to the set of features already selected in the previous steps $S \leftarrow S \cup \{k^*\}$ by maximizing the following equation:

$$k^* = \arg \max_{k \in F \backslash S} [J(X_k, Y | X_S) - \lambda c(k)], \tag{1}$$

where $J(X_k, Y | X_S) = I(Y, X_k) + \sum_{j \in S} [I(X_k, Y | X_j) - I(X_k, Y)]$ measures the informativeness of the added feature X_k in the context of already selected features and $\lambda c(k)$ is the penalty for the cost of the added feature. The $I(Y, X_k)$ is the mutual information metric and it measures the dependence between Y and X_k, $I(X_k, Y | X_j)$ is the conditional mutual information, whereas $I(X_k, Y | X_j) - I(X_k, Y)$ is the interaction information that measures the interaction between X_k and X_j in predicting Y. The general idea behind Eq. (1) is that it allows us to find the trade-off between the relevance of the feature and its costs and λ controls the balance. The optimal parameter λ value can be calculated by minimizing the loss function with the cross-validation on the training set.

3   Preliminary Results

The main goal of the experiments was to compare the cost-constrained algorithm for feature selection and to compare it with its traditional counterpart. By a traditional algorithm, we mean the method recalled in Eq. 1 with $\lambda = 0$. First, the parameter λ was optimized separately for each budget. Then we run a traditional and cost-constrained feature selection algorithm for $\lambda = 0$ and $\lambda = \lambda_{opt}$ respectively. Finally, we scored the Random Forest [2] model trained on the selected features with the AUC score. The optimization process, feature selection and model training were launched on the training data (50%) and the AUC metric is calculated on the test dataset (50%).

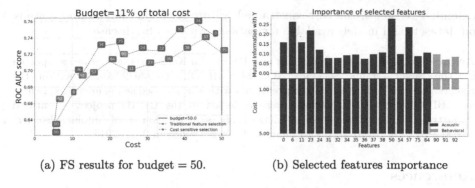

(a) FS results for budget = 50.

(b) Selected features importance

Fig. 1. Feature selection results.

Figure 1a presents the AUC of a model trained on features selected and in Fig. 1b we can see the importance of all selected features measured as mutual information with the target variable. Variables in the acoustic group result in greater mutual information with the target class, but at the same time, they are very expensive. On the other hand, we have behavioral features, which are 5 times cheaper, but their mutual information with the target class is lower. In the first step of both algorithms, the feature number 50 (magnitude of fast Fourier transform coefficients in band 250–650 Hz) from the acoustic group is selected. That is obvious because it is the most informative feature. In the second step cost-constrained method chooses feature number 92 (ratio of outgoing calls to all mobile calls per day) from the behavioral group and the traditional method feature number 6 (another magnitude of fast Fourier transform coefficients) from the acoustic group. The cost-constrained method in the third step selects feature number 36 (energy in the specific band), which combined with two previous features results in 0.70 AUC score, at the same time the traditional method results in 0.67 for almost the same budget. As the budget increases, both methods select more features, but the cost-constrained method tends to result in a higher AUC score. The cost-constrained method continues to be slightly better for classifying BD episodes than the traditional method.

4   Conclusions and Future Plans

The preliminary results conducted in this paper have shown that using the cost-constrained method let us select features that yield a more accurate classification model when restrictions on the budget are imposed in the BD episode classification problem. The difference in predictive power can be seen especially when the budget is low, for higher budgets both traditional and cost-constrained methods tend to equalize.

The current study points out that the implementation of only a small part of acoustic features (which are time/cost related for researchers) could be sufficient to receive first-worth results. The next steps for that study will be related to an extended version of the cost-constrained feature selection method and a study on datasets and models available for patients with mental disease.

Acknowledgment. Olga Kamińska and Katarzyna Kaczmarek-Majer are supported by the Small Grants Scheme (NOR/SGS/BIPOLAR/0239/2020-00) within the research project: "Bipolar disorder prediction with sensor-based semi-supervised Learning (BIPOLAR)". BDMON data were collected in the CHAD project − entitled "Smartphone-based diagnostics of phase changes in the course of bipolar disorder" (RPMA.01.02.00-14-5706/16-00) financed from EU funds in 2017–2018.

References

1. Bolón-Canedo, V., Porto-Díaz, I., Sánchez-Maroño, N., Alonso-Betanzos, A.: A framework for cost-based feature selection. Pattern Recogn. **47**(7), 2481–2489 (2014)
2. Breiman, L.: Random forests. Mach. Learn. **45**, 5–32 (2001)
3. Brown, G., Pocock, A., Zhao, M.J., Luján, M.: Conditional likelihood maximisation: a unifying framework for information theoretic feature selection. J. Mach. Learn. Res. **13**(2), 27–66 (2012). http://jmlr.org/papers/v13/brown12a.html
4. Dominiak, M., et al.: Behavioural data collected from smartphones in the assessment of depressive and manic symptoms for bipolar disorder patients: prospective observational study. J. Med. Internet Res. (2021)
5. Eyben, F., Weninger, F., Gross, F., Schuller, B.: Recent developments in opensmile, the Munich open-source multimedia feature extractor. In: Proceedings of the 21st ACM International Conference on Multimedia, pp. 835–838 (2013)
6. Grünerbl, A., Muaremi, A., Osmani, V.: Smartphone-based recognition of states and state changes in bipolar disorder patients. IEEE J. Biomed. Health Inform. **19**(1), 140–148 (2015)
7. Jagdhuber, R., Lang, M., Stenzl, A., Neuhaus, J., Rahnenführer, J.: Cost-constrained feature selection in binary classification: adaptations for greedy forward selection and genetic algorithms. BMC Bioinform. **21**(1), 1–21 (2020)
8. Kamińska, O., et al.: Self-organizing maps using acoustic features for prediction of state change in bipolar disorder. In: Marcos, M., et al. (eds.) KR4HC/TEAAM -2019. LNCS (LNAI), vol. 11979, pp. 148–160. Springer, Cham (2019). https://doi.org/10.1007/978-3-030-37446-4_12
9. Kamińska, O., Kaczmarek-Majer, K., Hryniewicz, O.: Acoustic feature selection with fuzzy clustering, self organizing maps and psychiatric assessments. In: Lesot, M.-J., et al. (eds.) IPMU 2020. CCIS, vol. 1237, pp. 342–355. Springer, Cham (2020). https://doi.org/10.1007/978-3-030-50146-4_26
10. Remeseiro, B., Bolon-Canedo, V.: A review of feature selection methods in medical applications. Comput. Biol. Med. **112**, 103375 (2019)
11. Teisseyre, P., Klonecki, T.: Controlling costs in feature selection: information theoretic approach. In: Paszynski, M., Kranzlmüller, D., Krzhizhanovskaya, V.V., Dongarra, J.J., Sloot, P.M.A. (eds.) ICCS 2021. LNCS, vol. 12743, pp. 483–496. Springer, Cham (2021). https://doi.org/10.1007/978-3-030-77964-1_37

ProbExplainer: A Library for Unified Explainability of Probabilistic Models and an Application in Interneuron Classification

Enrique Valero-Leal(✉) ⓘ, Pedro Larrañaga ⓘ, and Concha Bielza ⓘ

Departamento de Inteligencia Artificial, Universidad Politécnica de Madrid,
Madrid, Spain
enrique.valero@upm.es

Abstract. There are a multiplicity of libraries that implement Bayesian networks and other probabilistic graphical models. However, none of them is dominant, making it hard for users to deploy these models and build over them new functionality, such as new explainability algorithms that are specific for probabilistic models. We provide a common user interface called `ProbExplainer` for all of them over which algorithms are implemented, leaving to the user the relatively simple task to wrap up the specific implementation. We apply this library in an interneuron classification problem, a domain that is characterised by little expert agreement on labelling. We seek to study feature relevance through MAP-independence, an explainability method for Bayesian networks. The different expert models agreed that bigger subsets of unobserved features tend to be more relevant, the expert models are divided by whether the columnarity of an interneuron is irrelevant and in general the probability of a new observation changing the classification of its scenario is relatively low.

Keywords: Bayesian networks · Computational neuroscience · Software · Explainability

1 Introduction

Explainable artificial intelligence (XAI) is currently a discipline of great interest. It is an increasingly popular approach to deal with performance and ethical problems in complex artificial intelligence (AI) models, like neural networks.

Instead of focusing on black-box models, another interesting approach is to further enhance explainability and interpretability of transparent models [18]. Accordingly, much of our research focuses on Bayesian network (BN) explainability and, given that they are probabilistic models, they allow for explaining concepts such as uncertainty or lack of evidence.

However, a problem that arises when trying to deploy these models and explain them is that there is not a standard or ubiquitous implementation of

© The Author(s), under exclusive license to Springer Nature Switzerland AG 2024
J. M. Juarez et al. (Eds.): XAI-Healthcare/PM4H 2023, CCIS 2020, pp. 41–51, 2024.
https://doi.org/10.1007/978-3-031-54303-6_5

BNs, or even probabilistic models in general. On the contrary, we can find very popular implementations of other classification and regression algorithms (`scikit-learn` library) and even for specific models such as neural networks (`Pytorch` or `Tensorflow` libraries). We believe that this situation further discourages researchers from advancing in the field of BN explainability.

We propose to tackle this difficulty by designing a common interface for the different BN libraries and to implement explainability algorithms using it. As a result, algorithms will run their behaviour equally independently of the specific implementation. We further generalise this idea and extend this unified interface to all probabilistic models.

The contribution of this work is two-fold: (1) we present a library that allows for a common implementation of explainability algorithms for probabilistic models and (2) we test the library with an interneuron classification problem using one of its algorithms, namely MAP-independence [11]. A visual summary of the contributions can be seen in Fig. 1.

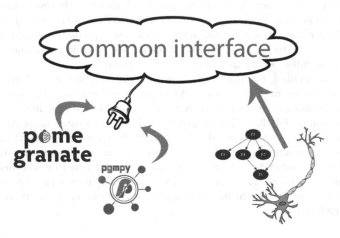

Fig. 1. Visual summary of the contributions of this paper: (1) an adapter to use any BN library and (2) an application in neuroscience

The paper is organised as follows: in Sect. 2, we provide the relevant background. In Sect. 3, we describe the architecture of our implementation and in Sect. 4, we apply it to a neuroscience problem. Finally, in Sect. 5, we draw some conclusions and elaborate on how to continue this research line.

2   Background

2.1   Probabilistic Models: Bayesian Networks

A (structured) probabilistic model [9] builds a joint probability distribution (JPD) over the random variables $\mathbf{X}$ of a problem. These models allow computing posterior probabilities of a set of variables $\mathbf{X}_j$ given another set $\mathbf{X}_i$ when the latter takes the joint value assignment $\mathbf{x}_i$, i.e., $P(\mathbf{X}_j \mid \mathbf{x}_i)$. In probabilistic graphical models, a graph is used to represent conditional independences that, in turn, allow us to represent the JPD into smaller but simpler factors. Examples of such models are Bayesian networks and Markov networks.

A Bayesian network $\mathcal{B} = (G, \theta)$ [9,15] represents a JPD $P(\mathbf{X})$ using a directed acyclic graph (DAG), G. The graph encodes conditional (in)dependence relations in $\mathbf{X}$ that allow to factorize the JPD as a product of the probability of each variable X given its parents in the graph, $\mathbf{Pa}_X$: $P(\mathbf{X}) = \prod_{X \in \mathbf{X}} P(X \mid \mathbf{Pa}_X)$.

An example of explainability for BNs used in this paper is MAP-independence [11], which studies the relevance of unobserved variables. Given a set of target variables $\mathbf{H}$, evidence $\mathbf{e}$ over the variables $\mathbf{E}$, a maximum a posteriori (MAP) $\mathbf{h}^* = \arg\max_{\mathbf{H}} P(\mathbf{H} \mid \mathbf{e})$ and a set of unobserved variables $\mathbf{R} \subseteq \mathbf{X} \setminus (\mathbf{H} \cup \mathbf{E})$, the result $\mathbf{h}^*$ is MAP-independent of $\mathbf{R}$ if $\forall \mathbf{r} \in \Omega(\mathbf{R})$ (value assignments of $\mathbf{R}$):

$$\arg\max_{\mathbf{h} \in \Omega(\mathbf{H})} P(\mathbf{h}, \mathbf{r} \mid \mathbf{e}) = \mathbf{h}^*.$$

MAP-independence was introduced as a method to check node relevance in BNs, that is, we say that a set of nodes $\mathbf{R}$ is irrelevant to a MAP result $\mathbf{h}^*$ if $\mathbf{h}^*$ is MAP-independent of $\mathbf{R}$. Put in another words, $\mathbf{R}$ is relevant to $\mathbf{h}^*$ if, in case it is observed, can alter $\mathbf{h}^*$.

The foundations of algorithms to find all the (ir)relevant subsets of unobserved features given a MAP query have also been presented [16]. Furthermore, the concept of MAP-independence strength [21] aims to quantitatively assess relevance, which is defined as the summation of the posterior probabilities $P(\mathbf{r} \mid \mathbf{e})$ of the value assignments $\mathbf{r}$ of $\mathbf{R}$ such that $\arg\max_{\mathbf{h} \in \Omega(\mathbf{H})} P(\mathbf{h}, \mathbf{r} \mid \mathbf{e}) = \mathbf{h}^*$.

In simpler words, irrelevance was defined quantitatively as the probability of $\mathbf{R}$ of not changing the value assignment $\mathbf{h}^*$ if observed jointly with evidence $\mathbf{e}$.

We believe that one of the greatest features of MAP-independence is to present a notion of feature (ir)relevance locally, i.e., when evidence $\mathbf{e}$ is present. Throughout this work, the term relevance for unobserved variables will be the introduced in this section using MAP-independence.

2.2   Existing Software

There are different open-source BN implementations, each one with its own characteristics, advantages and disadvantages. Some examples are:

- bnlearn [20] implements discrete and also Gaussian BNs and different learning algorithms. It is based on R, although heavy computations are made on C, improving its performance.
- bnclassify [14] offers a framework for learning and validating Bayesian classifiers in R, thus network structure is restricted.
- pgmpy [1] implements BNs in Python, using a relatively simple interface. Its biggest limitation is that it fully runs on Python, making it less efficient.
- pomegranate [19]. Similarly, this library implements BN networks in Python. The coding is done using Cython, decreasing the run time of algorithms.

– PyAgrum [6] is a wrapper of the library Agrum, which implements BNs in C. It is fast, includes many learning and inference algorithms, but the interface can be difficult to understand.

Regarding XAI libraries, to the best of our knowledge, there is none specialised in probabilistic (graphical) models. Most of them are usually dedicated to model-agnostic explainability (independent of the model used), focus on a single machine learning model or on a single XAI method.

2.3   Interneuron Classification: The Gardener Approach

Classification of GABAergic interneurons consists of labelling them given their morphological features, which are abundant and difficult to interpret.

A reliable classification is still an open challenge in neuroscience. These neurons have extremely diverse characterising features [2] and there is even no clear catalogue of existing types [4].

In [4], the authors propose to characterise interneurons by considering five simple axonal arborization descriptive features (which includes the type of interneuron) rather than their whole morphology. They seek to use a more utilitarian take on the problem referred to as "gardener approach", rather than the exact and scientific "botanic approach". To achieve this goal, they asked 42 experts to tag these five descriptive features for a representative set of interneurons of size 320. Although the classification was performed successfully for each individual expert, the tagging showed low consensus on axonal features among experts.

In later works [13], only interneurons whose label is backed by at least a certain number of experts are used. This results in a trade-off between the number of interneurons used for training classifiers and the reliability of the labels. They used different instances of Bayesian classifiers [3], which showed competitive performance for the problem. One of the models used was a tree augmented naive Bayes (TAN), which instead of assuming conditional independence of each feature given the class variables, allows for tree-like relations in the set of descriptive features.

Given the difficulties of the problem, such as the large number of morphological features, their relation to the five descriptive features and the different expert opinions, we believe that this is a problem well suited for an explainability study. In this work, we aim to locally study feature relevance for each expert.

3   Software Framework

Instead of designing another BN library on top of the existing ones, we provide a common interface to implement explainability algorithms. To this end, we identify queries and functions that can be used to design explanation algorithms, such as computing (joint) posterior probabilities or checking conditional independence (which can be easily done in probabilistic graphical models through d-separation [15]). Many existing explainability methods [5, 10, 11] only rely on these simple concepts. The final idea is to define algorithms over this common interface, disconnecting them from the actual implementation. An UML diagram can be seen in Fig. 2.

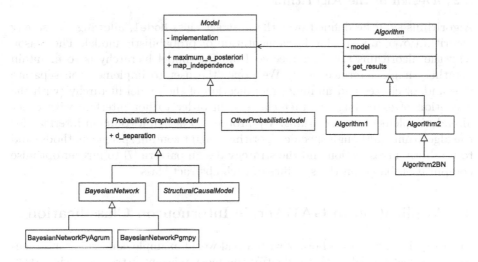

Fig. 2. UML diagram of the `ProbExplainer`

While the bulk of this paper focuses on BNs, other models such as structural causal models can be explained as well if adapted in the model hierarchy.

We decide to use Python for the implementation of our adapter. Adaption of R libraries is still possible using a language wrapper. The library is published in a public repository[1], alongside as the data and code necessary to replicate the experiments of Sect. 4.

3.1   A Unified Interface

To create a common interface with the methods of interest, we use an adapter design pattern [7]. We define an abstract class `Model` with an attribute containing the specific implementation of a probabilistic model and abstract methods representing queries that can be performed, such as `compute_posterior()`. More advanced functionality that relies on those abstract methods are also implemented here, such as computing MAP-independence. Other abstract classes inherit from `Model`, such as `ProbabilisticGraphicalModel`, creating an abstract hierarchy. These classes can extend the base functionality by adding new abstract methods.

Finally, to connect actual implementations, we need to create a specific class that inherits from the abstract model of interest, set the attribute

---

[1] https://github.com/Enrique-Val/ProbExplainer.

implementation with the model and override the abstract methods. As the bulk of the explainability algorithms are defined in other classes over the abstract classes, it is only necessary to override simple methods.

3.2 Design of the Algorithms

Algorithms shall be defined over the abstract class **Model**, allowing to use any algorithm over any specific implementation of probabilistic model. The reason why the algorithms are disconnected from the model hierarchy is to maintain a rather simple interface on it. We prefer the user to implement on separate files and/or classes and minimise modification of the model hierarchy (with the exception of extending abstract classes), in order to not interfere with other already implemented algorithms. We propose to design a common interface for the algorithms (although specific algorithms might contain specific methods) and to use both introspection and the strategy design pattern [7] to further optimise computation based on the specifics of each abstract class.

4 Application in GABAergic Interneuron Classification

In this section, we test a library with a real-world example concerning interneuron classification. We seek to identify the most relevant attributes using MAP-independence and related functionality. Unlike in previous works [13], we will not identify high reliability labels, but instead we will inspect the implicit expert opinion on feature relevance for each expert.

4.1 Data

The full dataset represents the characterisation in 5 features of 320 neurons made by each of the 42 experts. Full data was obtained using the R **gardenr** package[2]. It was processed on an R script to create 42 text files, each containing the annotation of the expert for all 320 interneurons. Although some additional data was dumped in said files, we only use the axonal features and drop the uncharacterised interneurons to remove noise. The dataset used is summarised in Table 1. For a more detailed description of the data, refer to [4] and [13].

4.2 Experiments

We decide to use the PyAgrum implementation of BNs and learn a TAN classifier. As such, we create the class **BayesianNetworkPyAgrum** that extends the abstract class **BayesianNetwork** and override the abstract methods. We build a TAN classifier for each of the experts. Although PyAgrum was selected as the test implementation, we learn the classifiers using **bnclassify** to keep consistency with previous works [13]. Each learned classifier will be converted into a

---

[2] https://github.com/ComputationalIntelligenceGroup/gardenr.

Table 1. Description of the dataset used

Variable	Labels
$F1$	intralaminar, translaminar
$F2$	intracolumnar, transcolumnar
$F3$	centered, displaced
$F4$	ascending, both, descending, none
$F5$ (class)	arcade, Cajal-Retzius, chandelier, common_basket, common_type, horse-tail, large_basket, Martinotti, neurogliaform, other

PyAgrum object. At the moment, PyAgrum is the only library adapted with `ProbExplainer`.

We use MAP-independence to measure unobserved feature relevance (as defined in Sect. 2) when one of them is observed. We assume that feature $F1$ is observed and its value is $F1 = intralaminar$. This will yield a MAP over the $F5$ class variable, which can differ for each expert. We aim to study if the MAP would change if additional evidence was found by each expert to study feature relevance. It is noteworthy that MAP-independence is a local method and, as such, relevance results yielded for the evidence $F1 = intralaminar$ may not be the same for other evidence.

We design two experiments for our work:

1. First, we use an algorithm to efficiently look for (ir)relevant variables using MAP-independence [16]. This experiment will determine the number of experts that consider each subset of features (ir)relevant. This is done indirectly using the BN learned from each expert, rather than directly consulting the expert directly.
2. Then we measure MAP-independence strength. Binarizing feature subsets in relevant/irrelevant may hide cases in which, for instance, a subset $\mathbf{R}$ is considered even with a single value assignment $\mathbf{r}$ with extremely low probability given the evidence $\mathbf{e}$ modifies the MAP result $\mathbf{h}^*$. Thus, we quantitatively assess the probability given the evidence $\mathbf{e}$ of a subset of features $\mathbf{R}$ of modifying the MAP result $\mathbf{h}^*$ if observed. Specifically, we use the MAP-independence strength complement to measure the probability of $\mathbf{R}$ modifying the result.

4.3 Results

The results of the first experiment are presented in Fig. 3. In general, experts consider higher-order subsets to be most relevant. In fact, most of them agree that if all the unobserved variables were observed, the result might have changed.

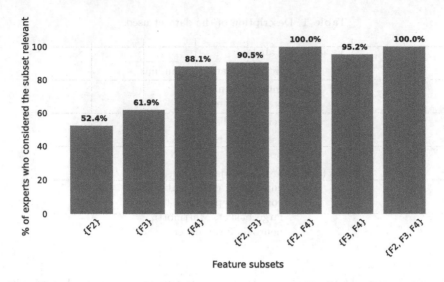

Fig. 3. Feature subset relevance

They did not identify an overwhelmingly irrelevant variable. Rather, we have cases such as the one of the variable $F2$ (which refers to the columnarity of an interneuron), which would make around half of the experts reconsider the most probable label (i.e. MAP) for $F5$, whereas the other half would still assign the same class to $F5$. If variables are considered individually, we can conclude that experts consider $F4$ to be the most relevant variable, while $F2$ the least, also being "polemical" as explained before.

The results of the second experiment can be seen in Fig. 4. We can note that most of the values stay within 0 and 0.5 probability ranges. We observe than on average large subsets still tend to be more relevant, similarly as in Fig. 3.

There are some details that slightly differ in both figures. Regarding singleton subsets, although in general $F4$ tends to be the most relevant, we can see that the probability of modifying $F5$ (given the evidence $F1 = intralaminar$) is actually lower if we observe $F4$ than if we observe $F2$ or $F3$. Analogously, many experts believed that $F2$ was irrelevant, but Fig. 4 shows that, if observed, the probability of modifying the MAP is higher.

In fact, we can consider $F2$ as a variable in which agreement of relevance is difficult. Roughly half of the experts consider it irrelevant. But the other half not only considers it relevant, but also believe that the value assignments that alter our MAP $\mathbf{h}^*$ are very likely given the evidence $\mathbf{e}$.

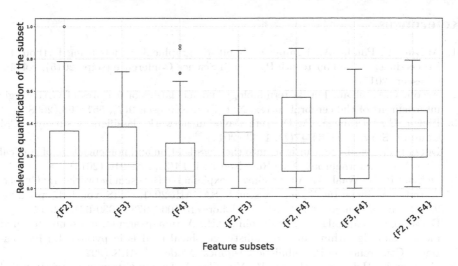

Fig. 4. Feature subset relevance quantification

5 Conclusions and Future Work

To summarise, ProbExplainer adapts concrete BN implementations, although a small loss in run time speed is expected. Regarding the application, the main finding is that in scenarios where the interneuron is intralaminar, the relevance of the columnarity on classification is a subject of debate between experts. As expected, large feature subsets are more likely to be considered relevant. Also, we identify cases in which a variable subset is not widely considered relevant, but its probability of actually modifying the output is higher than other relevant subsets.

In future work, we will further elaborate on how to implement algorithms in the library, as well as how to improve communications between the model hierarchy and the algorithm interface. It is of our interest to keep testing relevance through MAP-independence alongside existing methods, like LIME [17], SHAP [12] or counterfactuals [8], which offer a different view on feature relevance.

Acknowledgements. This research was supported by the Spanish Ministry of Science and Innovation through the PID2022-139977NB-I00 project and the TED2021-131310B-I00 "Bayesian Networks for Interpretable Machine Learning and Optimization (BAYES-INTERPRET)" project, and Ministry of Education through the University Professor Training (FPU) program fellowship (https://www.educacionyfp.gob.es/servicios-al-ciudadano/catalogo/general/99/998758/ficha/998758-informacion-comun.html), reference FPU21/04812.

References

1. Ankan, A., Panda, A.: Pgmpy: probabilistic graphical models using Python. In: Proceedings of the Fourteenth Python in Science Conference (scipy 2015), vol. 10. Citeseer (2015)
2. Ascoli, G.A., et al.: Petilla terminology: nomenclature of features of GABAergic interneurons of the cerebral cortex. Nat. Rev. Neurosci. 9(7), 557–568 (2008)
3. Bielza, C., Larranaga, P.: Discrete Bayesian network classifiers: a survey. ACM Comput. Surv. (CSUR) 47(1), 1–43 (2014)
4. DeFelipe, J., et al.: New insights into the classification and nomenclature of cortical GABAergic interneurons. Nat. Rev. Neurosci. 14(3), 202–216 (2013)
5. Derks, I.P., De Waal, A.: A taxonomy of explainable Bayesian networks. In: Gerber, A. (eds.) Artificial Intelligence Research. SACAIR 2021. CCIS, vol. 1342, pp. 220–235. Springer, Cham (2020). https://doi.org/10.1007/978-3-030-66151-9_14
6. Ducamp, G., Gonzales, C., Wuillemin, P.H.: Agrum/pyagrum: a toolbox to build models and algorithms for probabilistic graphical models in python. In: International Conference on Probabilistic Graphical Models. PMLR (2020)
7. Gamma, E., Helm, R., Johnson, R., Vlissides, J.: Design patterns: abstraction and reuse of object-oriented design. In: Nierstrasz, O.M. (eds.) ECOOP' 93 – Object-Oriented Programming. ECOOP 1993. LNCS, vol. 707, pp. 406–431. Springer, Berlin, Heidelberg (1993). https://doi.org/10.1007/3-540-47910-4_21
8. Guidotti, R.: Counterfactual explanations and how to find them: literature review and benchmarking. Data Min. Knowl. Discov. 1–55 (2022)
9. Koller, D., Friedman, N.: Probabilistic Graphical Models: Principles and Techniques. The MIT Press, Cambridge (2009)
10. Koopman, T., Renooij, S.: Persuasive contrastive explanations for Bayesian networks. In: Vejnarova, J., Wilson, N. (eds.) Symbolic and Quantitative Approaches to Reasoning with Uncertainty. ECSQARU 2021. LNCS, vol. 12897, pp. 229–242. Springer, Cham (2021). https://doi.org/10.1007/978-3-030-86772-0_17
11. Kwisthout, J.: Explainable AI using MAP-independence. In: Vejnarova, J., Wilson, N. (eds.) Symbolic and Quantitative Approaches to Reasoning with Uncertainty. ECSQARU 2021. LNCS, vol. 12897, pp. 243–254. Springer, Cham (2021). https://doi.org/10.1007/978-3-030-86772-0_18
12. Lundberg, S.M., Lee, S.I.: A unified approach to interpreting model predictions. Adv. Neural Inf. Process. Syst. 30 (2017)
13. Mihaljević, B., Benavides-Piccione, R., Bielza, C., DeFelipe, J., Larrañaga, P.: Bayesian network classifiers for categorizing cortical GABAergic interneurons. Neuroinformatics 13, 193–208 (2015)
14. Mihaljevic, B., Bielza Lozoya, M.C., Larrañaga Múgica, P.M.: Bnclassify: learning Bayesian network classifiers. R J. 10(2), 455–468 (2018)
15. Pearl, J.: Probabilistic Reasoning in Intelligent Systems. Networks of Plausible Inference. Morgan Kaufmann, Burlington (1988)
16. Renooij, S.: Relevance for robust Bayesian network MAP-explanations. In: Proceedings of the 11th International Conference on Probabilistic Graphical Models. Proceedings of Machine Learning Research, vol. 186, pp. 13–24 (2022)
17. Ribeiro, M.T., Singh, S., Guestrin, C.: Why should i trust you? Explaining the predictions of any classifier. In: Proceedings of the 22nd ACM SIGKDD International Conference on Knowledge Discovery and Data Mining, pp. 1135–1144 (2016)
18. Rudin, C.: Stop explaining black box machine learning models for high stakes decisions and use interpretable models instead. Nat. Mach. Intell. 1(5), 206–215 (2019)

19. Schreiber, J.: Pomegranate: fast and flexible probabilistic modeling in python. J. Mach. Learn. Res. **18**(1), 5992–5997 (2017)
20. Scutari, M.: Learning Bayesian networks with the bnlearn R package. J. Stat. Softw. **35**, 1–22 (2010)
21. Valero-Leal, E., Larrañaga, P., Bielza, C.: Extending MAP-independence for Bayesian network explainability. In: Proceedings of the Workshop Heterodox Methods for Interpretable and Efficient Artificial Intelligence. Zenodo (2022)

Interpreting Machine Learning Models for Survival Analysis: A Study of Cutaneous Melanoma Using the SEER Database

Carlos Hernández-Pérez[1] , Cristian Pachón-García[2] , Pedro Delicado[2] , and Verónica Vilaplana[1(✉)]

[1] Signal Theory and Communications Department, Universitat Politècnica de Catalunya.Barcelona Tech (UPC), Barcelona, Spain
{carlos.hernandez.p,veronica.vilaplana}@upc.edu
[2] Department of Statistics and Operations Research, Universitat Politècnica de Catalunya.Barcelona Tech (UPC), Barcelona, Spain
{cristian.pachon,pedro.delicado}@upc.edu

Abstract. In this study, we train and compare three types of machine learning algorithms for Survival Analysis: Random Survival Forest, Deep-Surv and DeepHit, using the SEER database to model cutaneous malignant melanoma. Additionally, we employ SurvLIMEpy library, a Python package designed to provide explainability for survival machine learning models, to analyse feature importance. The results demonstrate that machine learning algorithms outperform the Cox Proportional Hazards Model. Our work underscores the importance of explainability methods for interpreting black-box models and provides insights into important features related to melanoma prognosis.

Keywords: Survival Analysis · Machine Learning · eXplainable Artificial Intelligence · Melanoma

1 Introduction

The disruption of machine learning models has reached the field of Survival Analysis. In the last years, models such as Random Survival Forest (RSF) [4], DeepSurv [5] or DeepHit [7] have managed to achieve better accuracy, measured through the c-index [3], than the Cox Proportional Hazards Model [1] (CoxPH).

An example of this fact is a recent work [12] where the authors use a deep learning model which overtakes the CoxPH model, using the Surveillance, Epidemiology, and End Results database [8] (SEER) to model the cutaneous malignant melanoma. The authors use the CoxPH model, which provides a c-index of 0.875 and train a DeepSurv neural network which achieves a c-index of 0.91.

C. Hernández-Pérez and C. Pachón-García—These authors contributed equally to this work.

J. M. Juarez et al. (Eds.): XAI-Healthcare/PM4H 2023, CCIS 2020, pp. 52–61, 2024.
https://doi.org/10.1007/978-3-031-54303-6_6

One of the drawbacks of machine learning algorithms is their black-box nature, in the sense that the reasoning behind their predictions is often hidden from the user. However, in the last years some explainability methods have been developed to overcome this difficulty. Previously, we introduced SurvLIMEpy package [9], a Python library designed to obtain explainability for survival machine learning models. The underlying algorithm is SurvLIME [6], an agnostic local method inspired by LIME [11].

Given a survival machine learning model, $\hat{H} \colon \mathbb{R}^p \times \mathbb{R}_{>0} \to \mathbb{R}_{>0}$, and a test point $\mathbf{x}_* \in \mathbb{R}^p$, the goal of SurvLIME is to approximate the predicted value in a neighbourhood of $\mathbf{x}_*$ by the CoxPH model, $\hat{H}_{cox}(\mathbf{x}_*, t) = H_0(t) \exp(\hat{\boldsymbol{\beta}}^T \mathbf{x}_*)$. The vector $(\hat{\beta}_1 x_{*1}, \ldots, \hat{\beta}_p x_{*p})^T$ is used to rank the (local) importance of each feature.

In this work, we train three machine learning algorithms with the goal of finding insights about local feature importance. Therefore, we first train a RSF, a DeepSurv and a DeepHit model. Then, we compute feature importance for each of them. Since the CoxPH model is a standard in the Survival Analysis field, we also train this model. Then, we compare the results provided by SurvLIME algorithm with the coefficients of CoxPH model.

We ensure to obtain acceptable goodness-of-fit measures by means of using a 5-fold cross validation strategy to choose the values of the hyperparameters. The code used to produce this work is available at GitHub.[1]

2   Surveillance, Epidemiology, and End Results Database

The SEER database, launched in 1973, is composed by demographic, clinical and outcome data on all cancers diagnosed in representative geographic regions. The event indicator refers to the death and the time registered is either the time-to-event (when the individual eventually dies) or the time-to-censorship (the event is not observed), measured in months. More information about SEER is provided in [2,8].

2.1   Selection of the Individuals

Following the same selection criteria as in [12], we obtain the baseline dataset, which contains 34929 individuals. Table 1 describes all the criteria applied to obtain it. Then, it is randomly divided into two parts: the training dataset (24450 individuals; 70%) and the test dataset (10479 individuals; 30%).

The training dataset is used to model selection and training the algorithms. The test dataset is hidden to account for the final metrics. Since SEER database is updated periodically, the number of individuals may vary across data extraction processes (note that the number of individuals obtained in [12] is equal to 37758, where 26430 individuals are part of the training dataset and 11329 are part of the test dataset).

---

[1] https://github.com/CarlosHernandezP/xai-healthcare.

Table 1. Criteria applied to SEER database to obtain the baseline dataset.

Variable name	Value
ICD-O-3 Hist/behav	8720/3 Malignant melanoma, NOS
Primary Site - labeled	starts with 'C44'
date	from 2004 to 2015
Features used (listed in Table 2)	No missing values

Regarding the features, there are a total of 17 features, which are listed in Table 2. In Sect. 2.2, we analyse all of them using the training dataset. As specified in Table 1, the missing values are disregarded. Note that the age feature is provided as an ordinal feature in a yearly basis, with a total of 90 categories: starting at category '01 years' and ending at category '90+ years'. Therefore, all individuals older than 89 years belong to the same category. Due to this reason, we consider age as a categorical feature. The same happens with household income feature.

Table 2. Features used to model the data.

Name of the variables	Type
(1) tumor size, (2) tumor extension, (3) surgery of primary site	continuous
(4) age, (5) sex, (6) race, (7) marital status, (8) primary site, (9) T derived AJCC, (10) N derived AJCC, (11) M derived AJCC, (12) summary stage, (13) radiation status, (14) chemotherapy status, (15) lymph node dissection after surgery, (16) sequence of radiotherapy, (17) household income	categorical

2.2 Exploratory Data Analysis

In this section, we conduct an exploratory data analysis using the training dataset. After an initial analysis (not included in this work), we perform a dimensionality reduction by grouping the categories of the features. Concretely, two categories of a given feature are unified if the mean value of the event indicator is "similar". For instance, instead of dealing with the 90 categories of the age, we work with 3: 'age ≤ 44', 'age $\in (44, 64]$' and 'age > 64'.

The results are displayed in Table 3. The first column, named 'Total', accounts for the number of individuals per category, 'Num events' counts the number of individuals that have experienced the event, 'Rate events' is the mean value of the event indicator and 'Rate population' is the rate of population in each category. The mean value for the event indicator is equal to 0.32, with

Table 3. Results of the exploratory data analysis.

tumor size	Total	Num events	Rate events	Rate population
≤6	6755	1485	0.22	0.28
(6, 10]	7123	1833	0.26	0.29
(10, 16]	4551	1635	0.36	0.19
>16	6021	2836	0.47	0.25
tumor extension				
≤200	11983	3054	0.25	0.49
(200, 300]	5478	2191	0.40	0.22
>300	6989	2544	0.36	0.29
surgery of primary site				
≤30	6603	2358	0.36	0.27
(30, 31]	6097	1524	0.25	0.25
(31, 45]	8324	2740	0.33	0.34
>45	3426	1167	0.34	0.14
age				
≤44	3721	301	0.08	0.15
(44, 64]	9462	1546	0.16	0.39
>64	11267	5942	0.53	0.46
sex				
Female	9912	2338	0.24	0.41
Male	14538	5451	0.37	0.59
race				
other	941	167	0.18	0.04
white	23509	7622	0.32	0.96
marital status				
married	13473	4103	0.30	0.55
other	10977	3686	0.34	0.45
primary site				
other	19368	5483	0.28	0.79
first	5082	2306	0.45	0.21
T derived AJCC				
T1	16300	3853	0.24	0.67
other	8144	3936	0.48	0.33
N derived AJCC				
N0	21924	6380	0.29	0.90
other	2526	1409	0.56	0.10
M derived AJCC				
M0	23585	7256	0.31	0.96
other	865	533	0.62	0.04
summary stage				
localized	21664	6021	0.28	0.89
other	2786	1768	0.63	0.11
radiation status	Total	Num events	Rate events	Rate population
uncertain	24013	7449	0.31	0.98
other	437	340	0.78	0.02
chemotherapy status				
No/Unknown	24042	7506	0.31	0.98
Yes	408	283	0.69	0.02
lymph node dissection				
not	16464	5013	0.30	0.67
other	7986	2776	0.35	0.33

(*continued*)

Table 3. (*continued*)

tumor size	Total	Num events	Rate events	Rate population
sequence of radiotherapy				
no	24013	7449	0.31	0.98
yes	437	340	0.78	0.02
household income				
high	12308	3542	0.29	0.50
med	6219	2094	0.34	0.25
low	5923	2153	0.36	0.24

a standard deviation equal to 0.47. The mean value for the time-to-event (or time-to-censorship) is equal to 98.28 (months), with a standard deviation equal to 52.19.

3   Machine Learning Models

In this section, we explain the data preprocessing (Sect. 3.1) as well as the machine learning models we train (Sect. 3.2).

3.1   Data Preprocessing

As a first step, the training dataset is preprocessed. We transform categorical features into continuous ones using the target encoding technique [10]. Given a categorical feature, target encoding consist of using the mean value of the target in each of its categories. For instance, the age feature is encoded as follows: 0.08 if age $\leq$ 44, 0.16 if age $\in$ (44, 64] and 0.53 if age > 64. Note that the values used correspond to the ones obtained in 'Rate events' column of Table 3. The only feature without target encoding is sex, where we use one-hot-encoding: 1 for female and 0 for male.

We have decided to use this strategy because some works [10] show an improvement in terms of predictive capacity when target encoding is used. Additionally, we have performed a 5-fold cross validation process with two scenarios: (1) using the target encoding strategy and (2) using the traditional one-hot encoding strategy and we have obtained better metrics using (1). Regarding continuous features, we only transform tumor size by taking logarithms.

Once the features are transformed, we standardise all the 17 features so that each of them has a mean value equal to 0 and a standard deviation value equal to 1.

3.2   Machine Learning Models

We train three different machine learning models: RSF, DeepSurv and DeepHit. Additionally, we train a CoxPH model. We use a 5-fold cross validation strategy in order to select the best hyperparameters values. Afterwards, the test dataset is used to compute the c-index metric. The values obtained are 0.787, 0.798 0.796 and 0.802 for the CoxPH, RSF, DeepSurv and DeepHit respectively. In [12], the c-index of CoxPH and DeepSurv are better than ours, being 0.875 and 0.91 respectively. The difference could be explained due to the variation of the data between both data extraction processes. Nonetheless, our goal is not to obtain better performance but explain the machine learning algorithms.

In addition to the c-index values, the time-dependent AUC curves are computed for all the models. For a given threshold $\tau \in (t_{min}, t_{max})$, the event indicator is (re)computed as follows: if the individual experiences the event at time $t_{ind} \leq \tau$, then the (re)computed event indicator is equal to 1. In case the individual experiences the event later than τ or the individual does not experience the event, the (re)computed event indicator is equal to 0. Then, instead of dealing with a survival problem, the approach used is a binary classification problem. Thus, the AUC can be computed using as target the (re)computed event indicator and as predicted value the prediction of the risk at time τ. Figure 1 shows the time-dependant AUC curves for the four models, using the test dataset.

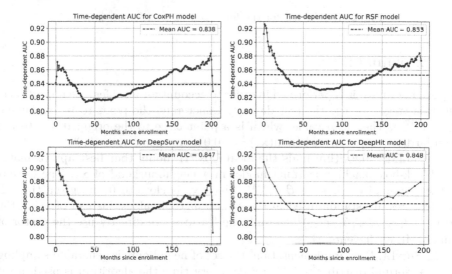

Fig. 1. Time-dependant AUC for the four models.

Finally, Fig. 2 depicts the ROC AUC curves at 1, 3 and 5 years for all the models. The survival problem is transformed into a binary classification problem using the same approach as the one explained previously. The test dataset is used to produce it.

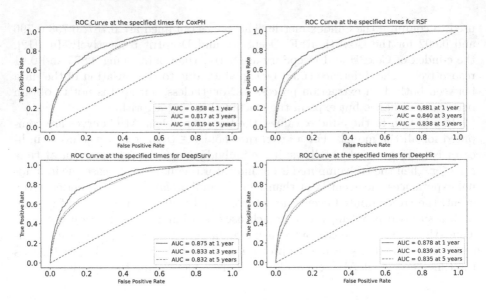

Fig. 2. ROC AUC for the four models at different times.

4   Explainability

A post-hoc analysis is performed in order to account for local feature importance. On this stage, we use the test dataset. Concretely, the point to be explained, $\mathbf{x}_* \in \mathbb{R}^{17}$, is the mean value, which is a point close to the origin $\mathbf{0}$ due to the standardisation performed ($\min(\mathbf{x}_*) = -0.015$ and $\max(\mathbf{x}_*) = 0.02$).

SurvLIME algorithm finds the local CoxPH model that best approximates the black-box machine learning model in a neighbourhood of $\mathbf{x}_*$. The output is the vector of coefficients $\hat{\boldsymbol{\beta}} \in \mathbb{R}^{17}$. Since $\mathbf{x}_*$ is close to $\mathbf{0}$ and given that all the features are standardised in the same range of values, instead of using the vector $(\hat{\beta}_1 x_{*1}, \ldots, \hat{\beta}_{17} x_{*17})^{\mathrm{T}}$ to rank the features, we directly use the vector of SurvLIME coefficients, $\hat{\boldsymbol{\beta}}$.

In SurvLIMEpy implementation, the set of neighbours is generated sampling from a Normal distribution. Therefore, every time the algorithm is used, a set of (new) neighbours is generated. We run 100 times the SurvLIME algorithm. Thus, we obtain a set of coefficients $\{\hat{\boldsymbol{\beta}}_1, \ldots, \hat{\boldsymbol{\beta}}_{100}\}$. We use those vectors to produce the figures included in this section.

Figure 3 upper-left depicts the feature importance according to SurvLIME when explaining the CoxPH model. According to SurvLIME coefficients, the most important feature for the CoxPH model is age: the older the individual is the higher its associated risk. With regards the sex feature, the risk associated with females is lower than for males. Note that we use colors, from red to blue, to display the boxenplots: in red are those features that increase the risk of experiencing the event and in blue those that decrease that risk.

Let $\bar{\beta}_{Surv}$ the vector containing the mean value of Fig. 3 upper-left. We can compare $\bar{\beta}_{Surv}$ with the vector containing the coefficients of the CoxPH model, $\hat{\beta}_{Cox}$. This comparison is depicted in Fig. 4. If SurvLIME algorithm estimated the CoxPH coefficients perfectly, then we would observe all the points in the straight line $y = x$ (depicted with a black dashed line). Although both sets of coefficients are not perfectly aligned, they remain comparably close.

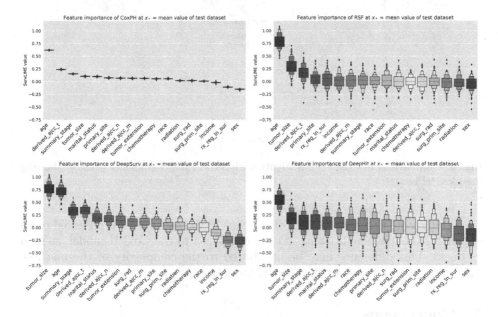

Fig. 3. SurvLIME coefficients for the four models.

Figure 3 upper-right, Fig. 3 bottom-left and Fig. 3 bottom-right represent the feature importance for RSF, DeepSurv and DeepHit respectively. The results suggest that the models exhibit similar feature importance, which can be interpreted akin to the CoxPH model. Notice how the models rely on the tumor size feature: the higher the tumor, the higher the risk. Although this feature is not as important as age, its role is very significant.

All the studied models attributed positive feature importance to the T, N and M derived AJCC (American Joint Committee on Cancer) features. The TNM system takes into account tumour size, lymph node involvement, and metastasis, which are globally accepted factors and provide a standardised way to describe the extent of cancer within a patient's body.

Additionally, it is observed that the order of the features vary across the models. This is due to the fact that each model has its own internal learning process and, as a result, feature importance can differ across algorithms.

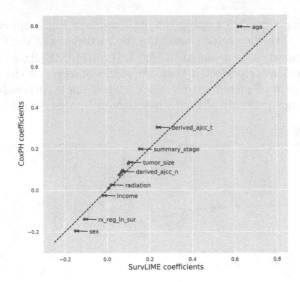

Fig. 4. Comparison of CoxPH coefficients with SurvLIME coefficients when explaining the CoxPH model.

Finally, note that the variance of the SurvLIME coefficients is observed to be higher for the machine learning models compared to the CoxPH model, as seen in Fig. 3. This difference may be attributed to the complexity of the machine learning models relative to the CoxPH models.

5   Conclusions

Machine learning methods are able to correctly model the time-to-death for cutaneous melanoma patients. On this study we focused on the interpretability of three black-box models using the SEER Database. On the one hand, we showed that machine learning algorithms can perform better than the CoxPH model. On the other hand, we computed local feature importance for a test individual, allowing to better understand the contribution of each feature, showing that the results obtained are consistent with clinically relevant features. This will help build trust between machine learning researchers and clinical practitioners.

Acknowledgments. This research was supported by the Spanish Research Agency (AEI) under projects PID2020-116294GB-I00 and PID2020-116907RB-I00 of the call MCIN/ AEI /10.13039/501100011033, the project 718/C/2019 funded by Fundació la Marató de TV3 and the grant 2020 FI SDUR 306 funded by AGAUR.

Disclosure of Interests. The authors have no competing interests to declare that are relevant to the content of this article.

References

1. Cox, D.R.: Regression models and life-tables. J. R. Stat. Soc. Ser. B (Methodol.) **34**(2), 187–202 (1972)
2. Duggan, M.A., Anderson, W.F., Altekruse, S., Penberthy, L., Sherman, M.E.: The surveillance, epidemiology, and end results (SEER) program and pathology: toward strengthening the critical relationship. Am. J. Surg. Pathol. **40**(12), e94–e102 (2016)
3. Harrell, F.E., Califf, R.M., Pryor, D.B., Lee, K.L., Rosati, R.A.: Evaluating the yield of medical tests. JAMA **247**(18), 2543–2546 (1982)
4. Ishwaran, H., Kogalur, U.B., Blackstone, E.H., Lauer, M.S.: Random survival forests. Ann. Appl. Stat. **2**(3), 841–860 (2008). https://doi.org/10.1214/08-AOAS169
5. Katzman, J., Shaham, U., Cloninger, A., Bates, J., Jiang, T., Kluger, Y.: Deepsurv: personalized treatment recommender system using a cox proportional hazards deep neural network. BMC Med. Res. Methodol. **18** (2018). https://doi.org/10.1186/s12874-018-0482-1
6. Kovalev, M.S., Utkin, L.V., Kasimov, E.M.: Survlime: a method for explaining machine learning survival models. Knowl.-Based Syst. **203**, 106164 (2020). https://doi.org/10.1016/j.knosys.2020.106164, https://www.sciencedirect.com/science/article/pii/S0950705120304044
7. Lee, C., Zame, W., Yoon, J., van der Schaar, M.: Deephit: a deep learning approach to survival analysis with competing risks. In: Proceedings of the AAAI Conference on Artificial Intelligence, vol. 32, no. 1, April 2018. https://doi.org/10.1609/aaai.v32i1.11842, https://ojs.aaai.org/index.php/AAAI/article/view/11842
8. National Cancer Institute, DCCPS, Surveillance Research Program: Surveillance, Epidemiology, and End Results (SEER) Program. https://www.seer.cancer.gov (April 2023), sEER*Stat Database: Incidence - SEER Research Data, 17 Registries, Nov 2022 Sub (2000-2020) - Linked To County Attributes - Time Dependent (1990-2021) Income/Rurality, 1969-2021 Counties, National Cancer Institute, DCCPS, Surveillance Research Program, released April 2023, based on the November 2022 submission
9. Pachón-García, C., Hernández-Pérez, C., Delicado, P., Vilaplana, V.: Survlimepy: a python package implementing survlime (2023)
10. Pargent, F., Pfisterer, F., Thomas, J., Bischl, B.: Regularized target encoding outperforms traditional methods in supervised machine learning with high cardinality features. Comput. Stat. **37**(5), 2671–2692 (2022)
11. Ribeiro, M.T., Singh, S., Guestrin, C.: Why should i trust you? Explaining the predictions of any classifier. In: Proceedings of the 22nd ACM SIGKDD International Conference on Knowledge Discovery and Data Mining, pp. 1135–1144. ACM (2016)
12. Yu, H., et al.: Deep-learning-based survival prediction of patients with cutaneous malignant melanoma. Front. Med. **10** (2023). https://doi.org/10.3389/fmed.2023.1165865, https://www.frontiersin.org/articles/10.3389/fmed.2023.1165865

Explanations of Symbolic Reasoning to Effect Patient Persuasion and Education

William Van Woensel[1]([✉])[iD], Floriano Scioscia[2][iD], Giuseppe Loseto[3][iD], Oshani Seneviratne[4][iD], Evan Patton[5][iD], and Samina Abidi[6][iD]

[1] University of Ottawa, 55 Laurier E., Ottawa, ON K1N 6N5, Canada
wvanwoen@uottawa.ca
[2] Polytechnic University of Bari, 70125 Bari, BA, Italy
[3] LUM "Giuseppe Degennaro" University, Casamassima, BA 70010, Italy
[4] Rensselaer Polytechnic Institute, Troy, NY 12180, USA
[5] Massachusetts Institute of Technology, Cambridge, MA 02143, USA
[6] Dalhousie University, Halifax, NS B3H 4R2, Canada

Abstract. Artificial Intelligence (AI) models can issue smart, context-sensitive recommendations to help patients self-manage their illnesses, including medication regimens, dietary habits, physical activity, and avoiding flare-ups. Instead of merely positing an "edict," the AI model can also explain *why* the recommendation was issued: why one should stay indoors (e.g., increased risk of flare-ups), why further calorie intake should be avoided (e.g., met the daily limit), or why the care provider should be contacted (e.g., prescription change). The goal of these explanations is to achieve understanding and persuasion effects, which, in turn, targets education and long-term behavior change. Symbolic AI models facilitate explanations as they are able to offer logical proofs of inferences (or recommendations) from which explanations can be generated. We implemented a modular framework called XAIN (eXplanations for AI in Notation3) to explain symbolic reasoning inferences in a trace-based, contrastive, and counterfactual way. We applied this framework to explain recommendations for Chronic Obstructive Pulmonary Disease (COPD) patients to avoid flare-ups. For evaluation, we propose a questionnaire that captures understanding, persuasion, education, and behavior change, together with traditional XAI metrics including fidelity (soundness, completeness) and interpretability (parsimony, clarity).

Keywords: Explainable Decision Support · Patient
Self-Management · Semantic Web

1 Introduction

Effective patient self-management relies on education of the health effects of dietary, activity, and medication choices [7]; as well as long-term behavior change,

J. M. Juarez et al. (Eds.): XAI-Healthcare/PM4H 2023, CCIS 2020, pp. 62–71, 2024.
https://doi.org/10.1007/978-3-031-54303-6_7

modifying unhealthy lifestyles that contribute to chronic illness [12]. Social Cognitive Theory (SCT) stipulates that health behavior will not be changed without sufficient motivation [1]. Among others, such motivation is influenced by knowledge of the health risks and benefits of behaviors [1]. Thus, by explaining the risks and benefits of health-related choices using domain knowledge, we aim to effect behavior change as well as education. Recent work has utilized Situational Awareness (SA) to frame eXplainable Artificial Intelligence (XAI) in terms of a series of informational needs [8]. In a health recommendation setting, these needs can be concretized as (1) *perception*: explanation of the recommendation itself; (2) *comprehension*: explaining the rationale of the recommendation, and (3) *projection*: projecting an alternative recommendation onto its required facts (*changed output → input*), and, inversely, projecting a set of alternative facts onto their corresponding recommendation (*changed inputs → output*). We argue that trace-based, counterfactual, and contrastive explanations can realize these informational needs, described by Chari et al. [2] as follows: (i) *trace-based*, outlining the sequence of reasoning steps that led to the recommendation (*comprehension*); (ii) *contrastive*, identifying the facts needed for an alternative recommendation (*changed output → inputs*); and (iii) *counterfactual*, identifying recommendations that would be issued for a set of given facts (*changed inputs → output*).

We present a modular explanation framework called XAIN (eXplanations for AI in N3) that generates (i) trace-based, (ii) contrastive, and (iii) counterfactual explanations. In this setup, a symbolic AI model is used to issue recommendations, with the model incorporating (a) a set of Notation3 (N3) [11] rules that encode recommendation logic; (b) a Personal Health Knowledge Graph (PHKG) [9], comprising the patient's health profile and environmental context; and (c) a domain Knowledge Graph (KG) with educational material. Based on logical proofs generated by the symbolic AI model, XAIN will generate human-readable explanations of the recommendation. In a bottom-up approach, XAIN generates descriptions of terms and tuples, including domain concepts (e.g., phlegm coloration) and recommendations (e.g., avoid going outside); aggregates the underlying reasons for a given inference or recommendation; and packages these inferences and their reasons into a presentation unit (e.g., HTML). The AI model and derivation proof is pre-processed to support different types of explanations. Other modules can be plugged in for additional types of explanations (e.g., scientific, contextual [2]) or customizing the presentation. XAIN expands our initial EXPLAIN framework [15] that supported trace-based explanations with 2 styles (visual and narrative). XAIN is open-source and available on GitHub[1].

This paper is structured as follows. In Sect. 2, we summarize and exemplify N3 derivation proofs. In Sect. 3, we introduce the XAIN framework and shortly summarize its modules. Section 4 demonstrates the XAIN explanations for a Chronic Obstructive Pulmonary Disease (COPD) use case. Section 5 outlines our planned evaluation. Finally, Sect. 6 ends with conclusions and future work.

---

[1] XAIN Framework, https://github.com/william-vw/xain.

2   Derivation Proofs

A derivation proof outlines the steps involved in inferring a statement. A step can involve (the extraction of) an asserted statement, the result of a built-in operation (e.g., sum), or the inference of another supporting statement. We utilize derivation proofs from N3-based symbolic AI models as supplied by the state-of-the-art *eye* [14] reasoner. The *eye* reasoner generates derivation proofs using the vocabulary presented by Verborgh, Arndt et al. (Definition 21, p. 26 [13]). The RDFS vocabulary can be found online[2].

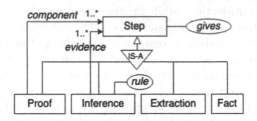

Fig. 1. Reason Vocabulary: relevant elements

Figure 1 summarizes the class structure and properties relevant to explanation generation. We shortly describe the different elements below:

- **Step**: individual step in a proof.
  - *gives*: graph (conjunction of statements) yielded by the proof step.
- **Proof**: inference to be proven (in our case, a recommendation).
  - *component*: proof *step* as one of the components of the proof.
- **Inference**: inference yielded by modus ponens (i.e., inference from a rule).
  - *rule*: logical implication (rule) yielding the inference.
  - *evidence*: list of *steps* that collectively entail the rule condition.
- **Extraction**: statement from a conjunction (e.g., domain KG).
- **Fact**: calculated results of built-ins (e.g., sum, concatenation)

Below, we show two rules inferring a recommendation to avoid going outside (referer an "environment recommendation")[3]. Next, based on these rules, we show a simplified derivation proof.

Listing 1.1. N3 rules for a recommendation based on the environment context.

```
1 @prefix : <http://example.org/>.
2 @prefix math: <http://www.w3.org/2000/10/swap/math#> .
3
4 { ?env :humidity :high } #rule1
5     => { ?env :recommendation :avoidGoingOutside } .
6 { ?env :humidity_value ?v . ?v math:greaterThan 50 } #rule2
7     => { ?env :humidity :high } .
```

---

[2] The Reason Vocabulary, https://www.w3.org/2000/10/swap/reason#.
[3] For the full ruleset, we refer to XAIN GitHub repository.

Listing 1.2. N3 proof for a recommendation based on the environment context.

```
1  @prefix : <http://example.org/>.
2  @prefix r: <http://www.w3.org/2000/10/swap/reason#>.
3
4  :proof a r:Proof;
5     r:gives { :currentEnv :recommendation :avoidGoingOutside } .
6     r:component :lemma1 .
7  :lemma1 a r:Inference ;
8     r:gives { :currentEnv :recommendation :avoidGoingOutside } ;
9     r:evidence ( :lemma2 ) ;
10    r:rule :rule1 . # rule1 (see Listing 1.1)
11 :lemma2 a r:Inference ;
12    r:gives { :currentEnv :humidity :high } ;
13    r:evidence (
14       [ a r:Extraction; r:gives { :currentEnv :humidity_value 55 } ]
15       [ a r:Fact ; r:gives { 55 math:greaterThan 50 } ]
16    ) ;
17    r:rule :rule2 . # rule2 (see Listing 1.1)
```

The component for the main *proof* is provided by rule-based inference *lemma1* (first rule, Listing 1.1), i.e., *:currentEnv :recommendation :avoidGoingOutside*, which is proven by step *lemma2* (second rule, Listing 1.1), i.e., *:currentEnv :humidity :high*. This step is itself proven by the assertion (extraction) that *currentEnv* has *humidity_value* 55, and the fact that 55 is greater than 50.

3 Human-Readable Explanation Generation from Derivation Proofs

XAIN is a modular framework for generating explanations from derivation proofs by an N3-based symbolic AI model. N3 (Notation3) is a Semantic Web (SW) rule language that supports Scoped Negation As Failure (SNAF), many built-in operations, and quoted graphs [11]. By using an SW language, the AI model can be easily integrated with a semantic PHKG and domain KG. Moreover, XAIN itself is implemented using N3, meaning it can directly operate on the N3-based AI model, its proofs, and semantic KGs, without any impedance mismatch. We used the N3 Visual Code plugin[4] to develop the XAIN framework.

As shown in Fig. 2, the symbolic AI model comprises an N3 recommendation ruleset (e.g., Listing 1.1) and is integrated with the patient's PHKG and domain KG. The AI model is executed by the *eye* reasoner, possibly resulting in a recommendation and derivation proof (e.g., Listing 1.2).

To generate a human-readable explanation, the XAIN **describe** and **collect** modules operate on the *derivation proof*, possibly obtaining labels and descriptions from the *domain KG*, to generate a single, human-readable presentation unit (e.g., HTML page). To support different types of explanations (e.g., contrastive), the **pre-process** module modifies the *domain ruleset* and resulting

---

[4] Van Woensel, W., Duval, T.: N3 Language and Reasoning Support, https:// marketplace.visualstudio.com/items?itemName=w3cn3.n3-lang-exec.

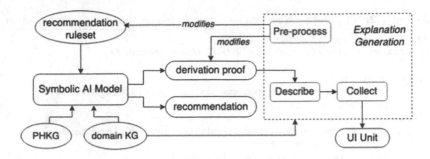

Fig. 2. XAIN Framework: Overall Process

derivation proof. We shortly summarize these modules below. For details, we refer to their implementation found in the GitHub repository.

3.1   Pre-processing Module

The *pre-processing* module will modify N3 artifacts to suit the requirements of different types of explanations:

Trace-Based and Counterfactual: Aggregate All Applicable Reasons. A trace-based explanation outlines the sequence of steps leading to the recommendation, starting from the current set of facts (e.g., from PHKG). We implement a counterfactual explanation in the same way; the only difference being that the facts can be dynamically set by the user. In both cases, a logical reasoner, starting from the facts, will follow a single sequence of steps to the recommendation. Additional rules can be added to Listing 1.1 that also recommend not going outside (e.g., based on temperature), and these may also hold based on the given set of facts. However, the proof only requires one such rule to prove the recommendation. While this makes sense from a logical standpoint, in our patient education case, it can be quite informative to list *all currently applicable* reasons for the recommendation; even if humidity was low or moderate, temperature or wind force could still result in a recommendation to avoid going outside[5].

   To that end, the pre-processing module (a) differentiates all rules with identical recommendations (e.g., environment recommendation)[6] to ensure the reasoner will treat them as separate proof components; (b) afterward, the module merges the *evidence* of all proof components that share the same original recommendation. For instance, in Listing 1.2 on line 9, this would result in other lemmas, one for each additional reason, being added to the list of evidence.

Contrastive: Inverting the Ruleset and Proof. A contrastive explanation identifies the facts required for an alternative recommendation (e.g., going outside). In other words, the starting point is not the facts (e.g., humidity, temperature) ultimately leading to a recommendation, but the recommendation itself.

---

[5] I.e., we are interested in the disjunction of reasons for the recommendation.

[6] This is done by adding random padding to the rule heads (i.e., consequents).

In line with this view, we implement a contrastive explanation as a trace-based explanation, but showing *all potential reasons* why a particular recommendation could be made. To that end, the pre-processing module first inverts the recommendation ruleset so that rule consequents become rule antecedents and vice-versa. For instance, an inversion of Listing 1.1 means the environment recommendation will infer a high humidity, which, in turn, will infer a (skolemized) humidity value that is greater than 50. After reasoning, the derivation proof will reflect the inverted ruleset, with the *proof* components being the facts and lowest-down *extractions* (evidence) being the recommendation. The pre-processing module subsequently re-inverts the derivation proof, putting the recommendation as the main *proof* and the facts as the lowest-down *extractions*.

3.2   Describe Module

The *describe* module generates human-readable descriptions for individual terms, based on their literal value, URI, or domain KG label. For compound terms (lists and graphs), strings of their constituent terms are combined into a single string.

As opposed to a comprehensive multi-level system based on external templates, as implemented by Dragoni et al. [3], this module is currently based solely on the N3 rules and domain KG. For instance, either the label provided by the domain KG is used, or "_" separated (or camel-case) URI local names are divided into their constituent words. A straightforward set of heuristics describe individual triples (e.g., "patient stratified yellow_zone") by printing the predicate and object, whereby a 'false' object adds a 'not' keyword ('true' object is ignored); in case of *evidence* that include comparisons (e.g., *lemma2* in Listing 1.2), strings are generated of the form "[predicate] [comparator] [value]".

3.3   Collect Module

The *collect* module collects the human-readable descriptions of a given *inference* (e.g., recommendation or high humidity) and its *evidences*, and packages them into a presentation unit; currently, HTML is supported. Inference results are wrapped into an HTML document that can be easily shown through a (mobile) Web browser. Explanation details are structured using a `header` element, including its direct evidence into a list of "evidence elements", e.g., using `div` and `p` tags designed to describe a paragraph of content within a Web document. To link these presentation units, a hyperlink is generated from an "evidence element" to its own associated presentation unit (e.g., high humidity). The *collect* module will obtain icons and summaries of domain concepts (e.g., humidity, temperature) from the domain KG and inject them into these HTML elements.

4   Demonstration

A live demo of the XAIN framework using Web technology can be found online. This demo relies on the *eye-js* N3 reasoner as it can be deployed in the browser[7] Fig. 3 reports two screenshots of a trace-based explanation type, referring to our Chronic Obstructive Pulmonary Disease (COPD) use case. The AI model issues a recommendation to the user to avoid going outside (Fig. 3a) due to high humidity and high wind. As shown in Fig. 3b, the user can select a specific reason listed in the trace-based explanation to get: (i) more details about low-level numeric data associated with the reason; (ii) a brief description summarizing the impact of the specific observation on COPD patients. Hence, educational material from the domain KG is also inserted into the presentation unit (Sect. 3.3). We point out that a contrastive explanation would be displayed similarly, with the alternative recommendation at the top, and all potential reasons listed in the unit body (see the online demo).

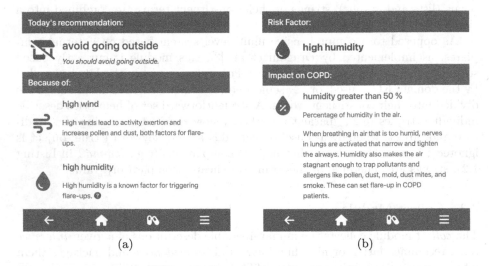

(a)   (b)

Fig. 3. Trace-based Explanation of Environment Recommendation.

Figure 4 shows our demo interface for counterfactual explanation types. A user is able to select any combination of facts to describe a hypothetical situation, check which recommendation would be generated, and view its explanation. In this case, the given facts do not indicate a situation that could trigger a COPD flare-up, so the AI model infers a recommendation to go outside. As before, the user can select any of the listed reasons to get its associated details.

---

[7] As *eye* is written in Prolog, it can be deployed in JavaScript using SWI-Prolog WebAssembly, https://www.swi-prolog.org/pldoc/man?section=wasm.

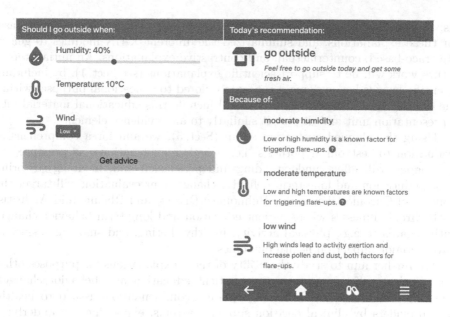

Fig. 4. Counterfactual Explanation of Environment Recommendation

5   Planned Evaluation

To evaluate an explanation approach, van der Waa et al. [10] propose to define the intended purpose of explanations in terms of measurable constructs; causal relations between these constructs constitute the hypothesis that is being evaluated. In our case, constructs include: (a) an understanding of the rationale behind the recommendation, (b) the persuasiveness of the explanation, (c) education on the health condition, and (d) behavior change towards healthy decisions. Our hypothesis involves the following causal relations: explanations will effect understanding and persuasiveness, which, in turn, will bring about education and behavior change. Confounding factors that could bias the measurement of these constructs include usability and perceived accuracy, which could respectively reduce the understanding of recommendations, and lead to an a priori bias towards trust [10]. At the same time, we aim to capture more traditional XAI evaluation metrics, including fidelity (soundness and completeness) and interpretability (clarity and parsimony). To capture the confounding factors and these evaluation metrics, we incorporate parts of PSSUQ [5] and Explanation Satisfaction Scale [4]. The resulting questionnaire, which will be used in our evaluation, can be found in our online Github repository.

6   Conclusions and Future Work

We presented an initial version of XAIN, a modular framework to generate human-readable explanations of recommendations issued by symbolic AI models. We outlined the structure of derivation proofs, which are used as a basis for these explanations, and summarized the different XAIN modules to generate trace-based, counterfactual, and contrastive explanations. An initial step in future work will be to support scientific explanations (see Sect. 1), by including scientific knowledge on domain concepts tailored to educational purposes within the domain KG. The *collect* module will include this educational material into a presentation unit and display it similarly to an "evidence element".

Using the planned evaluation setup (Sect. 5), we aim for a comprehensive evaluation to test our hypothesis: i.e., whether explanations, as presented in this paper, will affect understanding and persuasion and, in doing so, bring about education and long-term behavior change. Our evaluation will target the domain of Chronic Obstructive Pulmonary Disease and Rheumatoid Arthritis, both chronic illnesses where patient education and long-term behavior change, with regards to e.g., physical activity, healthy dieting, and smoking cessation, have a major impact on health outcomes.

We further aim to study the utility of these explanations for purposes other than patient self-management (i.e., patient education and behavior change). For instance, it may be useful to explain recommendations issued to health-care providers by clinical decision support systems; explaining the underlying reasoning steps, identifying the facts required for alternative recommendations, and which recommendations would fit a set of alternative facts. For that purpose, we will plugin other presentation types, such as summarized narrative text, that may be more suitable for these types of users. Next, we will evaluate our explanation approach in this context, using an adapted version of our questionnaire.

Deployment of the XAIN framework based on Web technologies will enable not only the creation of stand-alone (mobile) Web applications, but also integration into low-code development platforms like *Punya* [6]. This will facilitate rapid prototyping and development of native mobile apps by health researchers and practitioners that are able to explain their health recommendations.

References

1. Bandura, A.: Social cognitive theory of self-regulation. Org. Behav. Hum. Decis. Process. **50**(2), 248–287 (1991)
2. Chari, S., Seneviratne, O., Gruen, D., Foreman, M., Das, A., Mcguinness, D.: Explanation ontology: a model of explanations for user-centered AI. In: Pan, J.Z., et al. (eds.) The Semantic Web – ISWC 2020. ISWC 2020. LNCS, vol. 12507, pp. 228–243. Springer, Cham (2020). https://doi.org/10.1007/978-3-030-62466-8_15
3. Dragoni, M., Donadello, I., Eccher, C.: Explainable AI meets persuasiveness: translating reasoning results into behavioral change advice. Artif. Intell. Med. **105**, 101840 (2020)
4. Hoffman, R.R., Mueller, S.T., Klein, G., Litman, J.: Metrics for explainable AI: challenges and prospects. CoRR **abs/1812.04608** (2018). http://arxiv.org/abs/1812.04608
5. Lewis, J.R.: Psychometric evaluation of the post-study system usability questionnaire: the PSSUQ. In: Proceedings of the Human Factors Society Annual Meeting, vol. 36, no. 16, pp. 1259–1260 (1992)

6. Patton, E.W., Van Woensel, W., Seneviratne, O., Loseto, G., Scioscia, F., Kagal, L.: The Punya platform: building mobile research apps with linked data and semantic features. In: Hotho, A., et al. (eds.) The Semantic Web – ISWC 2021. ISWC 2021. LNCS, vol. 12922, pp. 563–579. Springer, Cham (2021). https://doi.org/10.1007/978-3-030-88361-4_33
7. Rose-Davis, B., Van Woensel, W., Raza Abidi, S., Stringer, E., Sibte Raza Abidi, S.: Semantic knowledge modeling and evaluation of argument theory to develop dialogue based patient education systems for chronic disease self-management. Int. J. Med. Inform. (2022)
8. Sanneman, L., Shah, J.A.: A situation awareness-based framework for design and evaluation of explainable AI. In: Calvaresi, D., Najjar, A., Winikoff, M., Framling, K. (eds.) Explainable, Transparent Autonomous Agents and Multi-Agent Systems. EXTRAAMAS 2020. LNCS, vol. 12175, pp. 94–110. Springer, Cham (2020). https://doi.org/10.1007/978-3-030-51924-7_6
9. Shirai, S., Seneviratne, O., McGuinness, D.L.: Applying personal knowledge graphs to health. CoRR **abs/2104.07587** (2021). https://arxiv.org/abs/2104.07587
10. van der Waa, J., Nieuwburg, E., Cremers, A., Neerincx, M.: Evaluating XAI: a comparison of rule-based and example-based explanations. Artif. Intell. **291**, 103404 (2021)
11. Van Woensel, W., Arndt, D., Tomaszuk, D., Kellogg, G.: Notation3: community group report (2023). https://w3c.github.io/N3/spec/
12. Van Woensel, W., Baig, W.H., Abidi, S.S.R., Abidi, S.R.: A semantic web framework for behavioral user modeling and action planning for personalized behavior modification. In: 10th International Conference on Semantic Web Applications and Tools for Life Sciences. CEUR, Rome, Italy (2017)
13. Verborgh, R., et al.: The pragmatic proof: hypermedia API composition and execution. Theory Pract. Log. Program. **17**(1), 1–48 (2017)
14. Verborgh, R., De Roo, J.: Drawing conclusions from linked data on the web: the eye reasoner. IEEE Softw. **32**(3), 23–27 (2015)
15. Van Woensel, W., et al.: Explainable clinical decision support: towards patient-facing explanations for education and long-term behavior change. In: Michalowski, M., Abidi, S.S.R., Abidi, S. (eds.) Artificial Intelligence in Medicine. AIME 2022. LNCS, vol. 13263, pp. 57–62. Springer, Cham (2022). https://doi.org/10.1007/978-3-031-09342-5_6

International Workshop on Process Mining Applications for Healthcare

PMApp: An Interactive Process Mining Toolkit for Building Healthcare Dashboards

Gema Ibanez-Sanchez$^{(\boxtimes)}$ [iD], Carlos Fernandez-Llatas[iD], Zoe Valero-Ramon[iD], and Jose Luis Bayo-Monton[iD]

Universitat Politècnica de València, ITACA-SABIEN, Camino de Vera, sn,
46022 Valencia, Spain
{geibsan,cfllatas,zoevara,jobamon}@upv.edu.es
http://www.sabien.upv.es/

Abstract. Process Mining is an emerging technology used in a variety of application domains, including healthcare. Despite existing tools for process analysis, the healthcare sector lacks a comprehensive solution addressing flexibility, connectivity, and usability. This paper presents the PMApp, the Interactive Process Mining Toolkit, a specialized tool for healthcare professionals. PMApp is designed to adapt to diverse scenarios, seamlessly integrate with health organizations' legacy systems, and facilitate Interactive Process Mining methodology. This methodology involves healthcare professionals in co-creation sessions to define, validate, and analyze Interactive Process Indicators, combining process-based perspectives with Key Performance Indicators. PMApp's innovative approach enhances information comprehension at different levels, making it user-friendly for healthcare professionals. The toolkit has been successfully tested with over one million patients across more than 10 European hospitals, addressing diverse healthcare scenarios in Portugal, Spain, Sweden, and The Netherlands.

Keywords: Interactive Process Mining · Data Rodeos · Interactive Process Indicators · PMApp Dashboard · Runner

1 Introduction

Process Mining is an evolving discipline that employs various techniques to enhance experts' understanding of involved processes and to facilitate data-driven process improvement. While widely applied in various business domains, its adoption in healthcare remains limited, primarily confined to research case studies [21]. Healthcare processes are inherently intricate, and characterized by high variability due to the complex nature of patient profiles and the deviation from established guidelines by Healthcare Professionals (HCPs) relying on their knowledge and experience.

Several dedicated Process Mining applications, like ProM, Disco, and Celonis, have emerged to improve processes across different sectors. ProM, with its

J. M. Juarez et al. (Eds.): XAI-Healthcare/PM4H 2023, CCIS 2020, pp. 75–86, 2024.
https://doi.org/10.1007/978-3-031-54303-6_8

extensive plugin support, offers flexibility but may overwhelm HCPs due to its complexity. On the other hand, applications like Celonis and Disco provide a more user-friendly experience, emphasizing data extraction, performance analysis, and scalability, albeit with potential limitations in customization and flexibility [19,25].

Recognizing the need for tools tailored to healthcare professionals, the Process Mining Community [19] advocates for the development of methodologies aligned with the healthcare domain. The *Interactive Process Mining* methodology (IPM) addresses this by involving healthcare professionals in the creation of tailored processes - *Interactive Process Indicators* (IPI) during special sessions called *Interactive Process Mining Data Rodeos*. This co-creation approach enhances HCPs' understanding of the process, though a challenge lies in the availability of tools conducive to this collaborative model.

To address this challenge and ensure the successful implementation of *IPM*, a critical prerequisite is an application that is interoperable, customizable, expandable, and user-friendly. This paper introduces PMApp, the Interactive Process Mining Toolkit designed specifically to initiate *Data Rodeos*, facilitating the application of the *IPM* methodology in real-world scenarios and promoting its acceptance among HCPs in their daily practice. *PMApp* has undergone testing in various case studies, including cardiology [2,16], cancer [23], obesity [24], and stroke [14], across different European hospitals such as those in Spain, The Netherlands, Sweden, and Portugal.

2 Through an Interactive Process Mining Solution for Healthcare

Over the past two decades, global life expectancy has increased by more than 6 years, driven not only by population growth but also by a surge in age-related and chronic diseases. Effective management of these conditions is crucial to alleviate their impact on individuals and society, potentially reducing the reliance on costly treatments[1].

Addressing this challenge, Artificial Intelligence, particularly paradigms like Process Mining, plays a vital role in assessing and enhancing patient care processes [6]. One such paradigm, *IPM* [6], uniquely involves HCPs in the process learning method. Unlike traditional approaches, *IPM* not only discovers but co-creates processes with HCPs, ensuring the resultant *IPIs* are relevant, trustworthy, and readily accepted in their daily practice.

IPIs, are defined as *Process representations that can be used to understand or measure the characteristics or intensity of one fact or even to evaluate its evolution* [6]. *IPIs* are navigable models, combining Process Mining techniques with domain-specific Key Performance Indicators (KPIs) [20], offer a more continuous, interactive, and understandable representation of real healthcare processes. For instance, in emergency departments, *IPIs* can integrate KPIs to measure

---

[1] https://www.who.int/news-room/fact-sheets/detail/noncommunicable-diseases.

Quality of Care (QoC), such as the length of stay [13,22] or the number of hyperfrequenters [15]. This holistic approach, combining KPIs and process analysis, provides a deeper understanding and flexibility in identifying root causes, setting *IPIs* apart from standalone KPIs.

The *IPM* methodology involves *Interactive Process Data Rodeos*, being key for the methodology with the main objective to build the *IPI* [6]. A *Data Rodeo* is defined as *a highly coupled multidisciplinary interactive data analysis aimed at building process indicators that allow understanding, quantifying and qualifying processes and their changes in an objective, comprehensive and exploratory way* [6]. A *Data Rodeo* is intended for iteratively curating data, co-creating process indicators, and validating them with HCPs and IT experts. These *Data Rodeos*, performed by Interactive Process Miners (Process Miners), are integral to building IPIs and advancing the co-creation process.

The IPM methodology unfolds in three phases: *Preparation*, where the multidisciplinary team aligns and specifies research goals; *Research*, involving iterative *Data Rodeos* to decide on the *IPI*; and *Production*, where the finalized *IPI* undergoes analysis [6].

Figure 1 illustrates various *Data Rodeos* conducted during the *Research phase*, beginning with *understanding* the data and identifying process steps and timestamps. The *cleaning process* follows, involving interactive collaboration to rectify or eliminate data deviating from the standard process. Discarded data is then revisited to distinguish between *wrong data and outliers*, providing valuable insights. The represented process serves as the foundation, with additional information such as averages and medians derived from it. Finally, the incorporation of *indicators* and the use of Process Mining *enhancement* techniques enrich the perspectives for a comprehensive understanding, ensuring the utility of the developed *IPI*.

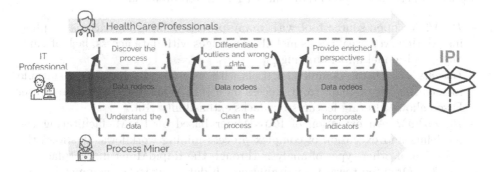

Fig. 1. Interactive Process Data Rodeo

However, *IPM* requires a specialized tool with distinctive features, including:

– **Flexibility.** The software should handle heterogeneous data sources seamlessly, offering tools for non-intrusive interconnection and supporting data transformation. It needs to be adaptable to specific use cases, allowing the creation of data pipelines for correction, transformation, and filtering. For instance, modifying an *IPI* for a new hospital's information system should be feasible.

- **Automatability.** An automatable tool should enable configuring and interpreting data processes as *IPIs*, making them repeatable with different data sources in the same format. It involves creating configuration files incrementally during the *Research* phase, allowing traceability and reuse in the *Production* phase.
- **Customizability.** The tool should incorporate new algorithms, techniques, and resources, allowing process miners, HCPs, or IT staff to enhance data analysis. This includes integrating new process mining techniques and varied ways of viewing information.
- **User-friendliness.** A user-friendly application is well-designed and easy to use, providing a positive experience for HCPs. This ensures acceptance and empowers HCPs to apply these techniques in their daily practice.
- **Understandability.** The tool should allow HCPs to comprehend the models generated by the system, adding value to their analysis. Additionally, it should facilitate the translation of *IPIs* into medical domain languages, such as GLIF, ProForma, Asbru or BPMN, making them understandable to experts.
- **Transparency.** The application should provide methods to assess the effects on data quality during data transformation and filtering. It should be transparent in each step when working with data, building trust by revealing what data is kept, and removed, and how indicators are calculated in the *IPI*.

According to previous systematic reviews [10,21] and other recent publications [4,5,12], Process Mining has been increasingly applied to a multitude of healthcare use cases since 2005. Several existing Process Mining tools have been applied in healthcare, potentially meeting *IPM* requirements:

- *ProM*. An open-source tool with many process mining algorithms as plugins [1]. However, weaknesses include challenges with data access, lack of guidance, and model understanding [3].
- *Celonis and Disco*. Commercial tools with user-friendly interfaces, suitable for creating visual maps from process data [11]. While they offer ease of use, Disco allows filtering features but lacks extensibility [10].
- *RapidProM*. An extension of RapidMiner based on ProM, facilitating the modeling and execution of complex process mining workflows. It allows the inclusion of other types of analysis through the RapidMiner marketplace.
- *UpFlux*. Oriented towards analyzing health data, but lacks customizability.

In summary, ProM is customizable but may not be intuitive for HCPs. Commercial tools like Disco are user-friendly but not customizable. RapidProM is automatable but lacks other requirements. UpFlux is health-data-oriented but falls short on various aspects. Notably, none of these tools offer the level of transparency described.

3   PMApp: An Interactive Process Mining Toolkit

This paper introduces *PMApp*, a dedicated desktop application designed to support the *IPM* paradigm in real-world scenarios, specifically aligning with the *Data Rodeo* process. *PMApp* addresses the challenges posed by data complexity, variability, and interoperability, providing a toolkit that allows the configuration of various Process Mining techniques and the customization of data views to create valuable *IPIs*. All of these configurations are consolidated into a single, user-friendly dashboard, ensuring ease of use for HCPs in their daily data analysis. Special attention is given to fostering trust among end-users in the application.

Built on the Microsoft .NET framework, *PMApp* is expressly tailored to support the *Data Rodeo* process during the *Research* phase of the *IPM* methodology. The toolkit boasts two key features, namely the *Experiment Designer* and *Ingestor Editor*, enabling high configurability to meet the specific needs of individual hospitals. This flexibility allows for the creation of custom dashboards, aligning with the unique requirements of health scenarios for daily practice in the *Production* phase.

As of now, *PMApp* stands out as the only toolkit accessible to .NET developers for creating Process Mining Research, catering to a segment of developers not covered by existing tools. Researchers can freely download *PMApp* for research purposes[2].

3.1   Experiment Designer

The *Experiment Designer* in *PMApp* is a module designed for creating formal schemes that define the Process Mining algorithm flow required to build *IPIs*. This module ensures the tool's *Automatability* by utilizing drag-and-drop blocks from the toolkit or developing new ones based on the healthcare organization's requirements. These blocks represent calls to functions and algorithms within the software to generate the *IPI*. The *Experiment Designer* is utilized during the *Research* phase by Process Miners, engaging in discussions with HCPs during *Data Rodeos* sessions. This collaborative approach defines the sequence of operations to transform hospital data into a final *IPI*. The Process Miner translates the *IPI* into an *Interactive Process Mining Runner* (*Runner*) as a configuration file automating the *IPI* creation process through drag-and-drop blocks defining the automation workflow by using the *Experiment Designer*.

The *Experiment Designer* feature is organized into five sections and facilitates the creation of *IPIs* through a user-friendly drag-and-drop interface. The process involves five key stages: log creation, filtering, log processing, TPA processing, and rendering.

– *Factories (1).* The first step involves creating logs using Factory blocks. Figure 2 is an example of synthetic data of an emergency department, where

---

[2] https://www.pm4health.com/download/.

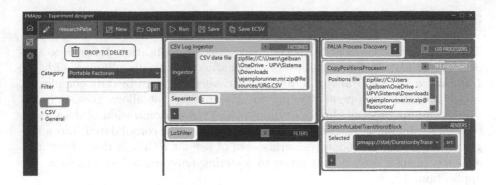

Fig. 2. Experiment Designer

can be identified a *CSV Log Ingestor* block, which manages data sources and establishes the rules to model the process in subsequent phases, being in this example reading a CSV file. Factories not only establish data connections but also provide data ingestion utilities for selecting data to create Events, Traces, and associated metadata. According to the *Data rodeo* phases, it is in charge of defining the raw log.

- *Filtering (2)*. After log creation, filters are applied to correct, split, and group traces, allowing modifications before the discovery process. Continuing with the example, blocks like *LoSFilter* (Length of Stay) model indicators available in the scientific literature and in *PMApp*. Likewise, the *Experiment Designer* allows the implementation of new blocks based on criteria provided by HCPs.
- *Log Processing (3)*. In the example (Fig. 2), the selected algorithm block is *PALIA Process Discovery* [7], which creates graphical models from logs. PMApp primarily uses Timed Parallel Automatons [8] (TPA) as the main formalism, being successfully tested in medical environments [17]. *PMApp* offers flexibility by translating TPA to other formalisms like BPMN or ProForma.
- *TPA Processing (4)*. In the TPA processing phase, TPA processors compute specific maps, statistics, or other post-discovery modifications. Blocks like *CopyPositionsProcessor* maintain the positions of nodes and transitions for future executions.
- *Rendering (5)*. In the final step, renders are applied to the graphical model (TPA) to highlight insights. Following the example, blocks like *StatsInfoLabelTrantitionsBlock* allow customization of the information displayed in transitions. In this case, the information shown is the average duration per trace, but other measures can be chosen before executing the *Runner*, which can be saved as a one-file-pack, simplifying execution on different computers, and heavy processing can be performed on supercomputers.

As Fig. 2 depicts, the *Experiment Designer*'s organizational structure aligns with the *Data Rodeo* flow, allowing Process Miners to configure *Runners* in a manageable way. Unique features include the ability to name, describe, and maintain a history of *Runners*, facilitating communication with HCPs. The extensibility of *PMApp* is evident through the addition of new blocks and resources, which is achieved by creating plugins in the .NET Core Framework, ensuring easy upgrades via the Settings menu.

3.2   Ingestor Editor

The *Ingestor Editor* (Fig. 3) is a vital module within *PMApp*, enabling the creation of logs from diverse data sources like CSV, SQL, etc. This module also facilitates the management of data quality. For instance, the opening of the *Ingestor Editor* reveals fields from the specified data source in block form, which can be incorporated into each ingestor step. The *Ingestor Editor* is organized into five tabs, with each tab serving a specific purpose:

- *Validators (A)*. This section features specialized algorithms that accept or reject rows during data ingestion based on criteria set by HCPs. Process Miners, in the example, use a block named *Line Rejector* to reject lines with null or empty triage start timestamps.
- *Variables (B)*. In this tab, blocks compute new variables that are combinations of existing data in the data table. These variables are often virtual fields, created by combining or extracting information from current fields. For instance, a *C# Operation* block can calculate a patient's age at the time of arrival by using C# code combined with CSV fields.
- *Events (C)*. The core of the ingestion process resides in this tab. Here, blocks are employed to create events using existing variables and fields. These events, representing nodes in the final model, require information such as a name, case ID, start, and sometimes end values. There are two types of blocks for defining events: *FieldEventExtractor*, which derives the event name directly from a selected field value (e.g., nodes for different discharge types), and *NamedEventExtractor*, allowing manual naming of events. Additionally, the *EventMetadataField* block adds metadata to events, such as triage level for *Triage* events.
- *Trace Data (D)*. This tab focuses on providing categorical information associated with each trace. This information, typically unchanging over time (e.g., gender), is crucial for stratification and characterization. Blocks in this tab contribute to statistics and domain-specific insights.
- *Filters (E)*. Similar to those applied in the *Experiment Designer*, the filters in this tab refine and enhance data during the ingestion process.

The *Ingestor Editor* plays a pivotal role in preparing and refining data, ensuring its quality and relevance for subsequent stages in the *PMApp* workflow.

3.3   Dashboard

The *Dashboard* serves as the primary module within the *PMApp*, providing a platform for visualizing and navigating the *IPI*. The *PMApp* functions as a container, assembling personalized *Dashboard* using the configuration information

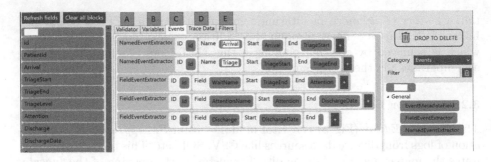

Fig. 3. Ingestor Editor

from the *Runner* file and integrating new resources through installed plugins (e.g., a calendar).

Upon completion of the *Dashboard* customization, HCPs and Process Miners can initiate the exploration of the *IPI*. Typically, the analysis commences with an examination of the *Main Perspective*, the central element of the *Dashboard* that illustrates the model with all the data (Fig. 4). Perspectives, representing diverse views of the data and models obtained through Process Mining techniques, are co-developed with Process Miners to enhance the user-friendliness and usability of the dashboard. These perspectives can be seamlessly added as plugins to the toolkit.

In the context of an emergency department scenario, HCPs may seek insights into the profile and characteristics of patients who exceed a 4-hour stay. Literature suggests an association between the length of stay and patient mortality rates [22]. Establishing a maximum value of 4 h [9] may consequently enhance service quality.

In Fig. 4, by clicking on the *@Start* node, the total number of emergency episodes in that year is 92,108. The colour intensity of nodes indicates the time patients spend on each activity. Notably, patients classified as level 1 (most urgent) may require more *Attention*. Moreover, the model can be enriched with domain-specific indicators, and measurable values assessing the process's performance in terms of efficacy and efficiency. The upper-right section of Fig. 4 illustrates the length-of-stay distribution, revealing that a significant number of patients exceed 4 h.

This information can be used in the *Group* menu, where different options can be combined. For instance, by combining the age and length of stay categories, a subprocess emerges, as shown in Fig. 5, indicating a total of 15,395 episodes. Multiple such combinations can be performed to yield various subprocesses as needed.

In Fig. 6, a detailed view of patient episodes passing through the *Attention1* node is presented. In this perspective, trace and event metadata are showcased, providing insights into patient profiles and the characteristics of their journey. Beyond the primary model, *additional perspectives* can be introduced to present information in various ways, enriching the understanding of the data.

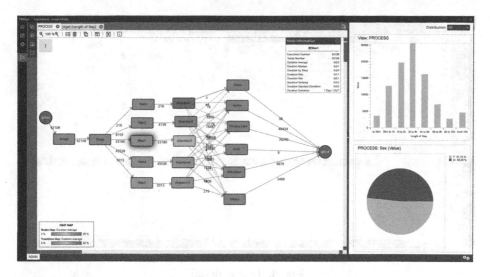

Fig. 4. IPI analysis - tab 1

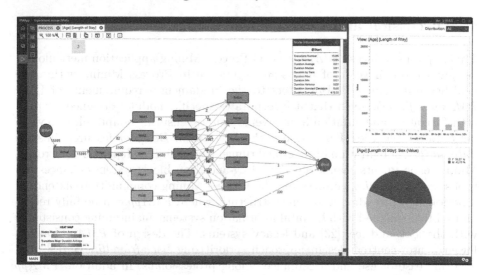

Fig. 5. IPI analysis - tab 2

For instance, a calendar perspective could depict the patient count per day, offering an alternative visualization for analysis purposes. Integrating such perspectives is seamless, requiring only the addition of a new plugin to the *PMApp*.

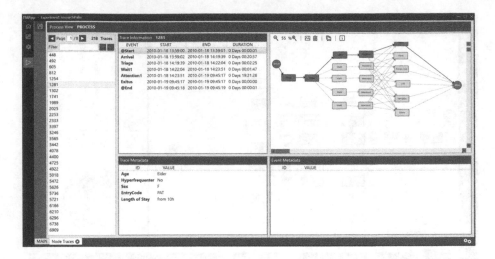

Fig. 6. Traces' detail

4   Discussion and Conclusions

This paper introduces *PMApp*, a novel Process Mining application meticulously designed through years of hands-on experience in Process Mining within the healthcare domain, *PMApp* caters to the fundamental requirements of IPM. *PMApp* is *flexible* given that it incorporates specific modules designed for data transformation and curation from diverse sources, ensuring adaptability to varying data complexities. Furthermore, the *Experiment Designer* feature, aims the *automation* configuration files, termed runners, enabling the repetition of Process Mining experiments seamlessly. The drag-and-drop block system enhances the tool's *customizability*, empowering the Process Mining community to extend its functionality, providing a modular structure, where *PMApp* can be fully reconfigured to be aligned with hospital information systems, maintaining consistency with the look and feel [23] and legacy systems. The design of *PMApp* has followed a user-centred design approach, prioritizing *user-friendliness* [18], and ensuring ease of use and acceptance among professionals. In addition, *PMApp* offers the ability to convert the main representation model (TPA) into others (BPMN, Proforma, etc.), enhancing *understandability* across different contexts. The toolkit provides perspectives for seamless navigation through structures (nodes and transitions) down to individual patients and events, ensuring *transparency* throughout the analysis.

Moreover, *PMApp* aligns with the Process Mining for Healthcare Manifesto [19] and provides an architecture based on the most common Process Mining methodologies but specifically designed to support the application of *IPM*. *PMApp* provides comprehensive tools, algorithms, and perspectives for information beyond discovery, (C1, C2). *PMApp* was designed and tested to tackle real-world data challenges (C4). It engages multidisciplinary teams, supporting

HCPs in utilizing Process Mining technologies effectively (C5). Likewise, specific blocks within *PMApp* address data quality concerns, providing ingestion reports to assess data quality and security (C6, C7). *PMApp* has been applied to diverse medical paradigms, including Evidence-based Medicine, Value-Based Healthcare, and dynamic risk models for chronic diseases (C10) [14].

Finally, *PMApp* has been successfully tested in multiple European use cases, including assessing the impact on ST-segment elevation myocardial infarction patients in Portugal [2], identifying bottlenecks in cardiology outpatient departments in Salamanca (Spain) [16], and analyzing oncological processes [23], dynamic risk models for chronic diseases [24], and Value-Based Healthcare approaches in various hospitals across Valencia (Spain) [14]. It has also been applied to emergency department processes in Stockholm (Sweden) and Rotterdam (The Netherlands) to reduce waiting times for frailty patients with stroke conditions.

References

1. Van der Aalst, W.M., van Dongen, B.F., Günther, C.W., Rozinat, A., Verbeek, E., Weijters, T.: ProM: the process mining toolkit. BPM (Demos) **489**(31), 2 (2009)
2. Borges-Rosa, J., et al.: Assessment of distance to primary percutaneous coronary intervention centres in ST-segment elevation myocardial infarction: overcoming inequalities with process mining tools. Digital Health **9**, 20552076221144210 (2023)
3. Claes, J., Poels, G.: Process mining and the ProM framework: an exploratory survey. In: La Rosa, M., Soffer, P. (eds.) BPM 2012. LNBIP, vol. 132, pp. 187–198. Springer, Heidelberg (2013). https://doi.org/10.1007/978-3-642-36285-9_19
4. Dallagassa, M.R., dos Santos Garcia, C., Scalabrin, E.E., Ioshii, S.O., Carvalho, D.R.: Opportunities and challenges for applying process mining in healthcare: a systematic mapping study. J. Ambient Intell. Hum. Comput. 1–18 (2022)
5. De Roock, E., Martin, N.: Process mining in healthcare-an updated perspective on the state of the art. J. Biomed. Inform. 103995 (2022)
6. Fernandez-Llatas, C.: Interactive Process Mining in Healthcare. Springer, Cham (2021)
7. Fernández-Llatas, C., Meneu, T., Benedi, J.M., Traver, V.: Activity-based process mining for clinical pathways computer aided design. In: 2010 Annual International Conference of the IEEE Engineering in Medicine and Biology, pp. 6178–6181. IEEE (2010)
8. Fernandez-Llatas, C., Pileggi, S.F., Traver, V., Benedi, J.M.: Timed parallel automaton: a mathematical tool for defining highly expressive formal workflows. In: 2011 Fifth Asia Modelling Symposium, pp. 56–61. IEEE (2011)
9. Geelhoed, G.C., de Klerk, N.H.: Emergency department overcrowding, mortality and the 4-hour rule in Western Australia. Med. J. Aust. **196**(2), 122–126 (2012)
10. Ghasemi, M., Amyot, D.: Process mining in healthcare: a systematised literature review. Int. J. Electron. Healthc. **9**(1), 60–88 (2016)
11. Günther, C.W., Rozinat, A.: Disco: discover your processes. BPM (Demos) **940**(1), 40–44 (2012)
12. Guzzo, A., Rullo, A., Vocaturo, E.: Process mining applications in the healthcare domain: a comprehensive review. Wiley Interdisc. Rev. Data Min. Knowl. Discov. **12**(2), e1442 (2022)

13. Horwitz, L.I., Green, J., Bradley, E.H.: US emergency department performance on wait time and length of visit. Ann. Emerg. Med. **55**(2), 133–141 (2010)
14. Ibanez-Sanchez, G., et al.: Toward value-based healthcare through interactive process mining in emergency rooms: the stroke case. Int. J. Environ. Res. Public Health **16**(10), 1783 (2019)
15. LaCalle, E.J., Rabin, E.J., Genes, N.G.: High-frequency users of emergency department care. J. Emerg. Med. **44**(6), 1167–1173 (2013)
16. Lull, J.J., et al.: Interactive process mining applied in a cardiology outpatient department. In: Munoz-Gama, J., Lu, X. (eds.) ICPM 2021. LNBIP, vol. 433, pp. 340–351. Springer, Cham (2022). https://doi.org/10.1007/978-3-030-98581-3_25
17. Martinez-Millana, A., Lizondo, A., Gatta, R., Traver, V., Fernandez-Llatas, C.: Expectations from a process mining dashboard in operating rooms with analytic hierarchy process. In: Daniel, F., Sheng, Q., Motahari, H. (eds.) BPM 2018. LNBIP, vol. 342, pp. 151–162. Springer, Cham (2019). https://doi.org/10.1007/978-3-030-11641-5_12
18. Martinez-Millana, A., Lizondo, A., Gatta, R., Vera, S., Salcedo, V.T., Fernandez-Llatas, C.: Process mining dashboard in operating rooms: analysis of staff expectations with analytic hierarchy process. Int. J. Environ. Res. Public Health **16**(2), 199 (2019)
19. Munoz-Gama, J., et al.: Process mining for healthcare: characteristics and challenges. J. Biomed. Inform. **127**, 103994 (2022)
20. Parmenter, D.: Key Performance Indicators: Developing, Implementing, and Using Winning KPIs. Wiley, Hoboken (2015)
21. Rojas, E., Munoz-Gama, J., Sepúlveda, M., Capurro, D.: Process mining in healthcare: a literature review. J. Biomed. Inform. **61**, 224–236 (2016)
22. Singer, A.J., Thode, H.C., Jr., Viccellio, P., Pines, J.M.: The association between length of emergency department boarding and mortality. Acad. Emerg. Med. **18**(12), 1324–1329 (2011)
23. Valero-Ramon, Z., et al.: Analytical exploratory tool for healthcare professionals to monitor cancer patients' progress. Front. Oncol. **12** (2022)
24. Valero-Ramon, Z., Fernandez-Llatas, C., Martinez-Millana, A., Traver, V.: Interactive process indicators for obesity modelling using process mining. In: Maglogiannis, I., Brahnam, S., Jain, L. (eds.) Advanced Computational Intelligence in Healthcare-7. SCI, vol. 891, pp. 45–64. Springer, Heidelberg (2020). https://doi.org/10.1007/978-3-662-61114-2_4
25. Van Der Aalst, W.: Process Mining: Data Science in Action, vol. 2. Springer, Cham (2016)

A Data-Driven Framework for Improving Clinical Managements of Severe Paralytic Ileus in ICU: From Path Discovery, Model Generation to Validation

Ruihua Guo[1,3] ⓘ, Ivy Sun[1], Crystal Chen[1], Qifan Chen[1], Yang Lu[1], Kevin Kuan[2], Abdulaziz Aljebreen[4], Owen Johnson[4], and Simon K. Poon[1](✉) ⓘ

[1] School of Computer Science, The University of Sydney, Sydney 2006, Australia
Simon.poon@sydney.edu.aus
[2] School of Information Systems and Technology Management, The University of New South Wales, Sydney 2052, Australia
[3] Australia Institute of Health and Welfare, Canberra 2617, Australia
[4] School of Computing, University of Leeds, Leeds 2006, UK

Abstract. Paralytic ileus (PI) is a severe health condition associated with poor clinical outcomes and longer hospital stays. Due to the high variability in clinical pathways, identifying risk factors on high-frequency pathways may facilitate the efficient optimization of clinical processes. This paper illustrated a data-driven framework that combines local process optimization and conceptual model validation. Frequent clinic pathways and contributing factors were discovered by leveraging local process modelling (LPM) and Partial Least Squares-based Structural Equation Modeling (PLS-SEM). Principle component analysis (PCA) was used to identify latent factors. LPM was used to identify structural relationships in the high-frequent process pathways. PLS-SEM was adopted to evaluate the magnitude of relations. Through this framework, the study identified one frequent clinic pathway and six contributing factors for severe PI patients.

Keywords: Process Mining · Factor Analysis · Partial Least Square

1 Introduction

Paralytic ileus (PI) is a severe clinical condition characterized by intestinal blockage and the absence of abdominal smooth muscle activation [1, 4]. In 2019, 10.1 million people worldwide were diagnosed with PI, resulting in over 7 million global disease burden measured by disability-adjusted life years [2]. Hospitalization for PI is particularly prevalent among people aged 65–79 years and is associated with increased cost of care [3]. The etiology of severe PI is multifaceted and lacks a precise mechanism [4]. The obstruction may arise from either mechanical or non-mechanical factors. Severe PI can be attributed to a variety of factors, including abdominal surgery, fluid imbalances, infections, and the use of analgesics and antidepressants [1]. Limited understanding of

J. M. Juarez et al. (Eds.): XAI-Healthcare/PM4H 2023, CCIS 2020, pp. 87–94, 2024.
https://doi.org/10.1007/978-3-031-54303-6_9

optimization of severe PI clinical pathway, especially severe patients in ICU, resulted in a poor clinical outcome of PI patients, including high mortality rate and prolonged hospitalization in untreated individuals [5].

Electronic health records (EHRs) present a unique opportunity for managing PI patients in clinics, as they can capture comprehensive information on clinic activities, diagnoses, clinical outcomes, and expenditures. By utilizing EHRs to identify frequently occurring clinical pathways in episode care and risk factors for severe PI hospitalization, healthcare providers can optimize resource allocation and improve the clinic outcomes of severe PI patients.

Process mining is a novel technique that employs data mining algorithms and process modelling to analyze event logs recorded by a health information system. Local process model mining, as opposed to exploring end-to-end models, focuses on capturing frequent local patterns. Structural equation modelling (SEM) is a scientific investigation technique that uses visualization and model validation to explain pre-assumed causal relationships among latent constructs [6].

The aim of this study is to build a data-driven framework for supporting dynamic clinical management of severe PI hospitalization and uncovering common contributing factors to frequent clinic processes.

2   Method

The framework was constructed in five phases: cohort extraction, event log building, local process discovery, conceptual model building, and strategy generation (see Fig. 1).

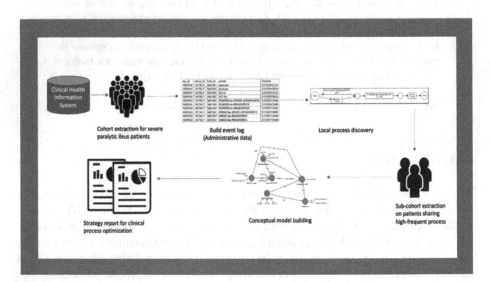

Fig. 1. Structure of a data-driven framework for optimizing clinic management in severe paralytic ileus patients.

2.1 Data Resource

MIMIC-IV version 2.0, released on June 12, 2022, is a large-scale publicly available healthcare database corresponding to over 60,000 patients admitted to the Beth Israel Deaconess Medical Center in Boston in the United States [7, 8].

2.2 Cohort Extraction

A cohort of patients with severe PI and their episode care was extracted following criteria: Patients aged 16 and older, diagnosed with PI coded as ICD-10 code K56.0. Admitted to ICU at least once, and with a medication record for Prokinetics, Opioid Antagonists or Laxatives.

This cohort was used to identify high-frequent local processes for patients with severe PI. After that, a sub-cohort containing severe PI patients who underwent the frequent local process was extracted from the original cohort. Hospital length of stay (LoS), defined as the length of time elapsed between a patient's hospital admittance and discharge, was calculated as an indicator of the clinical burden for severe PI patients. A longer LoS is indicative of a more complex treatment scenario and a greater burden for patients and healthcare providers.

2.3 Event Log Extraction

An event log refers to a collection of events, each with a timestamp that records the executed time. An event represents a unique execution of an activity, which is a well-defined step in the process, such as "laboratory test" [9]. A trace will represent the order of events, and all events in a single trace will be carried out by the same context, which is typically characterized in terms of patient treatment flows.

To extract the event log of the cohort, a quality-aware framework [10] was followed, and the extraction was further justified by domain knowledge. The resulting event log contains all the necessary therapy activities and their execution timestamps, including different types of abdominal surgery and medications for severe PI patients.

2.4 Frequent Patient Pathways Discovery

In contrast to end-to-end models, the process discovery through a local approach focuses solely on identifying local process models (LPMs) with a lower number of activities, usually between three and five [11]. The LPMs are then matched with the process tree to systematically investigate the frequent pathways, up to a predetermined model size [11]. For local process discovery, the LPMs developed by Tax et al. [11] were utilized, and the quality of the resulting local processes was assessed based on five metrics: *Confidence* measures the degree to which the event conforms to the LPM derived from the event log; *Determinism* indicates the level of predictability of future behaviors; *Language fit* is defined as the ratio of behaviors allowed by the LP observed in the event log; *Coverage* measures the frequency of events that can be identified in the log; *Score* evaluates the overall performance of the LPM based on the aforementioned metrics.

2.5   Structural Equation Modelling

We assumed that patients who follow the same clinic pathways would share common characteristics and risk factors. However, establishing the connections between multiple factors can prove to be a challenging task.

In this study, principal component analysis (PCA) was adopted to generate hypothesis and identify potential contributing factors. Partial Least Squares-based Structural Equation Modeling (PLS-SEM) is employed to understand the relationship among the identified factors, due to its effectiveness in handling multiple structural paths without any distributional assumptions on the data and accommodating both small and large sample sizes [12, 13].

3   Results

Total of 647 suitable severe PI patients' records were extracted from MIMIC-IV version 2.0 described in Table 1. On average, these patients were 63 years old, with an age range from 21 to 96 years. Of the 647 patients, 65% (423) were male, while 35% (224) were female. The higher proportion of males may be attributed to the fact that female PI patients have been reported to have a higher in-hospital mortality rate and lower incidence [14]. The average length of hospital stay for these patients was 20 days, with a range of 1 to 137 days.

Table 1. Demographic summary for 647 hospital-admitted severe PI patients (the original cohort).

Demographic summary	All subject, n = 647		
Sex	Males	Females	People
Count, N (%)	423 (65%)	224 (35%)	647 (100%)
Age in years, Mean (SD)	63 (14.97)	62 (17.07)	63 (15.71)
Length of Stay in days, Mean (SD)	19 (15.29)	22 (15.84)	20 (15.51)

A total of 7985 clinic event logs have been compiled, and the leading 10 frequently occurring events have been summarized in Fig. 2. Treatment with prokinetics was found to be the most common activity for severe PI patients, followed by laxative treatment. The frequency of these two clinic activities was higher than the time of admission and discharge, suggesting that many severe PI patients may have received more than one episode of care involving prokinetics and laxatives. In-hospital death was identified as the 9th most common event, accounting for 1.5% of the total events, for severe PI patients.

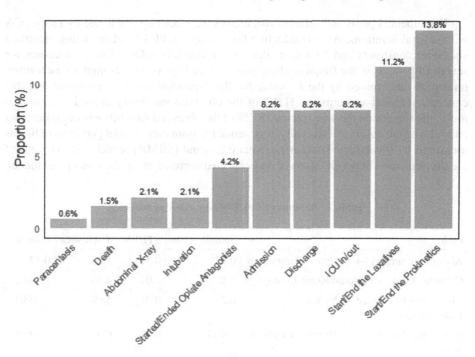

Fig. 2. Leading 10 frequent events for hospital-admitted severe PI patients.

Ten frequently occurring local processes have been identified through LPMs. Among these, we identified one local process that aligned with clinic practices and achieved the highest score (0.87) and best performance on the Confidence, Determinism, Language fit, and Coverage matrices (0.99, 1.0, 1.0, 0.24, respectively). Our findings suggest that, in severe PI hospitalizations, patients were typically transferred to the ICU after being admitted to the hospital, followed by discharge from the index hospital (as shown in Fig. 3).

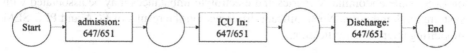

Fig. 3. Petri Net of showing high frequent severe PI patient pathway.

Following exclusion of in-hospital deaths, we identified 526 severe PI patients who followed the frequent pathway. Similar distribution of age and sex was detected in this sub-group when comparing to the full cohort. Of the 526 patients, 65% (344) were male, while 35% (182) were female. The average length of hospital stay for these patients was 20 days, with a range of 2 to 137 days.

Hypothetical paths and selected risk factors were built up and tested by using PCA with several iterations. After conducting bootstrapping, PLS-SEM modeling identified six latent constructs and 24 factors that may potentially affect clinical outcomes for severe PI patients in the frequent clinic process (see Fig. 4). The strength of each structural path, determined by the R^2 value for the dependent variable, indicated that this conceptual model accounts for 31.9% of the observed variability in the LoS. Furthermore, the regression model explains 13.2% of the observed variability in complications and 7.1% of the observed variability in systemic inflammation. Model goodness of fit was measured by standardized soot mean square residual (SRMR), which reflects 7.2% of the discrepancies between observed and expected correlations in the conceptual model.

Table 2. Summary of PLS-SEM modelling statistics.

Path	Sample mean	STDEV	T statistics	P values
Abdominal surgery → System Inflammation	0.42	0.27	1.55	0.12
Complications → Hospital length of Stay	0.42	0.01	4.95	<0.001
Electrolyte Imbalance → System Inflammation	0.25	0.04	5.60	<0.001
Electrolyte Imbalance → Hospital length of Stay	−0.21	0.04	4.97	<0.001
Metabolic Disturbance → Complications	0.17	0.02	8.71	<0.001
Metabolic Disturbance → Hospital length of Stay	0.18	0.05	3.97	<0.001
System Inflammation → Hospital length of Stay	−0.25	0.03	8.40	<0.001

Table 2 describes the model, including path coefficients and P values. Six components, with eigenvalues larger than 1, have been identified through PCA. According to the conceptual model, complications and metabolic disturbances have a significant impact on LoS, while abdominal surgeries and electrolyte imbalances may be associated with systemic inflammation. No significant difference on the results of PCA and PLS-SEM modelling after sex stratification.

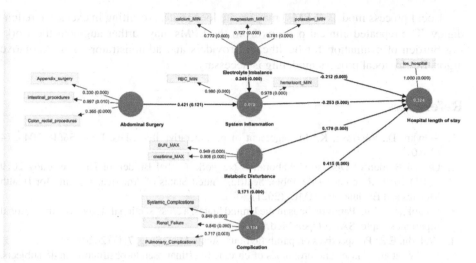

Fig. 4. Conceptual model for contributing factors in the frequent clinic pathway. Blue nodes represent constructed latent factors; Yellow cubes reflected including variables; Number on each path represented path coefficient (P values): P-value less than 0.01 indicated strong evidence.

4   Discussion and Conclusion

This study illustrated a data-driven framework, starting from pathway discovery and progressing to model building, followed by model testing. This framework is aimed at enhancing the clinical management of severe PI in the ICU, utilizing LPM and PLS-SEM techniques. Through PLS-SEM, both direct and indirect relationships have been uncovered along the pathways. It has been observed that metabolic disturbance has a direct impact on LoS and also exerts an indirect effect on LoS through the mediation 'Complications'. However, the relationship between electrolyte imbalance and systemic inflammation with LoS should be interpreted with caution. Based on current evidence [15, 16], there is no consistent relationship among the three constructs.

The low score of item-to-construct reliability may need further investigation. Investigation of combining domain knowledge for paths as well as valid features measuring the events that have direct or indirect associations to outcome will be future direction. As hypotheses generation tools with clinical and administrative domain experts for inputs to generate realistic hypotheses for statistical testing would significantly enhance the applicability of mined outcomes in clinical practice. Evidence derived from current research to support conceptual model building and evaluation would ensure the quality of proposed clinical strategies.

Nonetheless, the absence of information on procedure duration and medication dispensing time in this secondary data may impede our understanding of their roles in clinical processes for severe PI patients. Exploring datasets that encompass these clinical activities may uncover more accurate optimizing strategies. Moreover, the current framework mainly focuses on frequent clinic pathways, so patients who received uncommon treatments and their specific needs have not been fully discovered.

Local process models (LPMs) may create loop routes, resulting in excessive redundancy. The repeated clinical patterns from the LPMs may further augment the working burden of evaluation for healthcare providers and administrators. An alternative algorithm for local process modeling is necessary.

References

1. Stewart, D., Waxman, K.: Management of postoperative ileus. Dis. Mon. **56**(4), 204–214 (2010)
2. Global Burden of Disease Collaborative Network. Global Burden of Disease Study 2019 (GBD 2019) Reference Life Table. Seattle, United States of America: Institute for Health Metrics and Evaluation (IHME) (2021)
3. Solanki, S., et al.: Paralytic ileus in the United States: a cross-sectional study from the national inpatient sample. SAGE Open Med. (2020)
4. Weledji, E.P.: Perspectives on paralytic ileus. Acute Med. Surg. **7**(1) (2020)
5. Lu, W., et al.: Causes and prognosis of chronic intestinal pseudo-obstruction in 48 subjects. Medicine **97**(36), e12150 (2018)
6. Dash, G., Paul, J.: CB-SEM vs PLS-SEM methods for research in social sciences and technology forecasting. Technol. Forecast. Soc. Chang. **173**, 121092 (2021)
7. Johnson, A., Bulgarelli, L., Pollard, T., Horng, S., Celi, L.A., Mark, R.: MIMIC-IV (version 2.0). PhysioNet (2022)
8. Goldberger, A., et al.: PhysioBank, PhysioToolkit, and PhysioNet: components of a new research resource for complex physiologic signals. Circulation **101**(23), e215–e220 (2000)
9. Lu, Y., Chen, Q., Poon, S.: Detecting and understanding branching frequency changes in process models. In: Augusto, A., Gill, A., Nurcan, S., Reinhartz-Berger, I., Schmidt, R., Zdravkovic, J. (eds.) BPMDS/EMMSAD -2021. LNBIP, vol. 421, pp. 39–46. Springer, Cham (2021). https://doi.org/10.1007/978-3-030-79186-5_3
10. Andrews, R., van Dun, C.G., Wynn, M.T., Kratsch, W., Roglinger, M., ter Hofstede, A.H.: Quality-informed semi-automated event log generation for process mining. Decis. Support. Syst. **132**, 113265 (2020)
11. Tax, N., Sidorova, N., Haakma, R., van der Aalst, W.M.: Mining local process models. J. Innov. Digit. Ecosyst. **3**(2), 183–196 (2016)
12. Hair, J.F., Risher, J.J., Sarstedt, M., Ringle, C.M.: When to use and how to report the results of PLS-SEM. Eur. Bus. Rev. **31**(1), 2–24 (2019)
13. Hox, J.J., Bechger, T.M.: An introduction to structural equation modelling (1998)
14. Koşar, M.N., Görgülü, Ö.: Incidence and mortality results of intestinal obstruction in geriatric and adult patients: 10 years retrospective analysis. Turk J. Surg. **37**(4), 363–370 (2021)
15. Rehman, F.U., Omair, S.F., Memon, F., Amin, I., Rind, B.J., Aziz, S.: Electrolyte imbalance at admission does not predict the length of stay or mortality in dengue-infected patients. Cureus. **12**(9), e10419 (2020)
16. Jeon, C.Y., Neidell, M., Jia, H., Sinisi, M., Larson, E.: On the role of length of stay in healthcare-associated bloodstream infection. Infect. Control Hosp. Epidemiol. **33**(12), 1213–1218 (2012)

Phenotypes vs Processes: Understanding the Progression of Complications in Type 2 Diabetes. A Case Study

Roberto Tornero-Costa[1]([✉]) [ID], Antonio Martinez-Millana[1] [ID],
and Juan-Francisco Merino-Torres[2] [ID]

[1] Universitat Politècnica de València, 46022 Valencia, Spain
{rotorcos,anmarmil}@itaca.upv.es
[2] La Fe Health Research Institute, 46026 Valencia, Spain

Abstract. Patients with type 2 diabetes mellitus are at high risk of developing several types of complications. Previous studies show a clear association of tight glycemic control, both in terms of glycated hemoglobin and the time in range, with different metabolic pathways that contribute to nervous, micro, and macro-vascular complications. As a matter of fact, a few works attempted to obtain phenotyping to describe the evolution of type 2 diabetes patients; however, due to the complexity and variability of the disease progression, together with factors related to the process of care focused on medical specialties, made it difficult to obtain clinically meaningful processes to understand the progression of the disease. In this work, we apply process mining techniques to automatically detect temporally restricted patterns based on routinely collected clinical data to define possible pathways and disease progression process models to support clinicians in understanding the process of care in type 2 diabetes mellitus. Our work is focused on the most prevalent complications, cardiovascular and nephropathy, based on a retrospective cohort of 16,187 patients during eight years of follow-up and demonstrates the ability of process mining approaches to identify disease progression patterns.

Keywords: process discovery · decision support systems · type 2 diabetes mellitus · diabetes complications · process mining

1 Introduction

Process mining is a discipline that focuses on extracting meaningful information and knowledge from logs of events and aims to provide analytics on the executed processes that produced that data [1]. Type 2 diabetes is a chronic metabolic disorder that can develop multiple comorbidities. Persons with type 2 diabetes, which account for the 90% of persons with diabetes, are at an increased risk of cardiovascular and nervous complications [2]. The management of type 2 diabetes is based on the continuous refinement of therapies and lifestyle counseling (e.g.

J. M. Juarez et al. (Eds.): XAI-Healthcare/PM4H 2023, CCIS 2020, pp. 95–106, 2024.
https://doi.org/10.1007/978-3-031-54303-6_10

diet, physical activity) that aims to maintain and increase patient quality of life by preventing complications [3]. Clinical and non-clinical activities support this continuous process, and despite the knowledge about the condition, every case is different due to the multifactorial nature of the condition and the influence of the patient behavior and decisions of the clinical team (endocrinologist, nurses, educators...) [4].

Healthcare processes for managing type 2 diabetes are highly dynamic, complex, multidisciplinary, and usually fitted for purpose, in the way that each endocrinology unit has its own manner of implementing standard clinical protocols. Improving healthcare processes is not an easy task, even though it is clear that by its optimization, patients' quality of life and use of clinical resources can be increased and optimized [5]. Several strategies have been used to analyze hospital processes, including Business Process Redesign [6] and Evidence-Based Medicine [7]. Process Mining principles may be useful to obtain new perspectives on how patients are managed and which are the most common clinical pathways they follow across the provision of care process [8].

The economic impact of type 2 diabetes has two components [9]. The first involves social impact, quantified as the costs involved in the purchase of treatments, productivity decrease as means of lost labor days, disability, and mortality. The second is the healthcare impact, quantified as the economic costs involved in hospital admissions, outpatient services visits, primary care follow-up, and pharmacological and not pharmacological treatment and tests. As population life expectancy increases, type 2 diabetes prevalence will increase, driving an augmented impact on social and healthcare.

The main therapy goal of type 2 diabetes management consists of maintaining blood glucose levels as much of the time into the normality ranges. The glycated hemoglobin (A1C) is currently the gold standard indicator for the performance of the ongoing treatment. The literature has demonstrated that keeping adequate levels of A1C reduces the incidence of short and long terms complications such as micro/macrovascular and peripheral nervous deterioration. However, it is complex to link the relationship between different diagnoses and the overall disease progression [3].

In this study, we focus on the analysis of type 2 diabetes-related complications and their relationship in the short and long term. Process mining techniques were applied to discover which are the most common followed by the majority of patients on the two major groups of comorbidities: cardiovascular and kidney complications. Abstract models of the disease progression are necessary to properly evaluate the benefits and differences of different diabetes treatments and care pathways.

2   Methodology

2.1   Data Collection and Study Measures

The study consisted of a retrospective observational study of electronic health records of type 2 diabetes patients diagnosed between 2012–2019 at the health department of the University and Polytechnic Hospital La Fe in Valencia, Spain. Diagnostic criteria were based on ADA guidelines for Fasting Glucose Plasma ≥ 126 mg/dL, positive Oral Glucose Tolerance Test or A1C ≥ 6.5 mg/dL [10]. Hospital La Fe is a reference hospital covering a population of nearly 300,000 inhabitants, from which a total of 16,187 patients with type 2 diabetes were eligible for the study. Data were collected from the Data Warehouse of Hospital La Fe, which contains different sources of clinical information (diagnoses, laboratory tests, prescriptions, consultations, emergencies, hospital admissions, etc.).

In order to identify the most common pathways in the disease progression, the scope of the study was to reduce the population into different cohorts with similar demographic criteria based on the following: (1) patients with a type 2 diabetes disease progression of 3 to 6 years and (2) patients by age between $[0, 35)$, $[35, 60)$ and $[60,)$ years based on their age at the first related activity. The disease progression was evaluated as the period between the first HbA1c laboratory test or first diagnosed related complications to the last record in the retrospective study. For each patient, the last record may consist of the last registered diabetes-related complication or use of health resources (before the year 2020) or the closure of the electronic health record due to death.

2.2 Methods

The approach of this study was to consider all the events registered in the electronic health records of patients diagnosed with type 2 diabetes. We shall then consider all the transitions among these events an iterative process, which overlaps each chain of events into the same workflow. As a result, it is expected to obtain complex process models with hard-to-understand information. However, they may provide another type of information about the complete and complex scenario of the development of comorbidities. Prior to applying the process mining models, data were extracted from clinical data warehouses, pre-processed into the data corpus, and finally pre-processed with preliminary filters depending on the type of analysis.

The goal of the study then was not to implement new process mining techniques or algorithms but to apply current process mining techniques to discover new information in the corpus of data and use this information to answer specific clinical questions. The processing strategy pipeline we propose is made up of four sequential steps:

- First, we describe the data used for the process mining analysis by assessing the population and observation characteristics across the different datasets and age segments.
- Second, we perform a pattern discovery by attacking the datasets using the process discovery technique based only on the diagnosis codes for complications.
- Third, we compare the retrieved process model abstraction among the same segments of different populations.

– Fourth, we compare (when possible) results with other types of care flow analysis.

Diabetes-related diagnoses were categorized into six high-level groups: acute, cardiovascular, kidney, ophthalmic, neurological, and peripherical complications. These categories comprise a total of 47 different pathologies and complications, such as hyperglycemia, heart failure diagnoses, nephropathies, retinopathies, cataracts, etc. Often, the relation between type 2 diabetes progression and the use of resources by the patient is studied throughout the whole patient's history. The top common complications and diagnoses can be selected at a patient level by attending to their relative frequencies in the cohort. Also, the sequence of first occurrences of diabetes complications in each category can be studied in order to understand the most common pathways.

Disease progression process models can be defined as the time-based sequence of acute diagnoses or the first occurrence of chronic complications in a patient. Based on clinical guidelines, diabetes progression is evaluated in intervals of 6 months. Diagnoses, emergency admissions, outpatient visits, hospital admissions, HbA1c tests, and exitus of patients were processed into a log format and grouped over periods of 6 months. Therefore, for every patient, a set of traces was defined as a collection of records during a 6-month interval. Based on sequential rules, the relative frequency of diabetes complications in these traces can be studied to evaluate the most significant. For the most prevalent sequences (>5%), different analyses can be applied, such as the use of different health resources, the relation between distinct categories of diagnoses, differences between traces of patients with different complexity, or differences between traces where patients died or not. The analysis was implemented in R v4.2.2 [11], with the libraries tidyverse v2.0.0, dplyr v1.1.0, and bupaR v.0.5.2 [12].

2.3   Data Corpus and Event Log Generation

The type-2 diabetes data corpus is composed of (a) patients demographic data (gender, birth date, end of clinical history if applies), (b) diagnose registers (ICD-9 and ICD-10 diagnoses codes, dates), (c) laboratory tests (HbA1c, triglycerides, LDL, HDL, Creatinine analysis, results, dates), (d) emergency admissions (dates, severity and triage results, related diagnoses, outpatient derivations), (e) hospital admissions (admission date, discharge date, related diagnoses, discharge reason, related surgeries), (f) home hospitalization (admission date, discharge dates, related diagnoses, discharge reasons), (g) outpatient services visits (date appointments, type of service, completion or not of the visit), (h) primary care visits (date appointments, primary measures like weight, BMI or blood pressure, and their results), (i) other medical examinations (retinographies, electrocardiograms, kidney dialysis, and their dates), (j) indicators of patient's care estimations (Risk Adjustment Factors - RAF - and clinical risk group - CRG - categories, and dates the indicators were calculated).

A patient event log was generated based on the occurrence of diabetes-related diagnoses, exitus, and HbA1c laboratory analysis. Each patient corresponds to a case and each registration of diagnose or laboratory analysis is an activity. Emergencies, hospital admissions, outpatient visits, primary care visits, and medical examinations were integrated in the log as trace metadata. Outpatient appointments not completed were removed. Further, a second event log was generated, breaking cases into 6-month intervals to create 6-month traces to study the disease in the short term. In this case, only activities and metadata between the years 2012 and 2020 were considered. For each patient, the computation of 6-month periods was performed with the first HbA1c analysis or the first diabetes-related diagnosis as the first trace's starting point.

3 Results

3.1 Patient-Level Analysis

From the total population of N=16,187 patients, 6,622 patients met the inclusion criteria and were divided into 2 cohorts: patients aged 35 to 59 years and patients aged 60 years or more. The group of patients younger than 35 was discarded due to only including 21 samples. The cohort of oldest patients comprises 4,924 patients, and 3,217 patients presented diabetes-related complications. Also, 55.9% of the patients with complications, n = 1,798, have registered the end of the clinical history in the interval 2012–2019. The top common diagnoses affecting patients are those related to cardiovascular (63% of the patients) and kidney (32% of the patients), whereas the occurrences of acute (16%), peripheral (16%), ophthalmic (15%) and neuro (7%) complications are substantially lower.

The second cohort is integrated by 1,698 patients and 821 had complications. In this case, dropouts were reduced to only 20,5% of the patients, n = 169. Moreover, the distribution of diabetic complications changes. While cardiovascular complications are still the most common (44,58% of the patients), kidney complications are the category with the least prevalence (13,52%). Acute (22%), neuro(19%), ophthalmic (18%), and peripheral complications (14%) have a higher prevalence than kidney diseases in younger patients. Figure 1 presents the traces of these patients, where a trace is a set of occurrences of a unique group of diagnoses. While the top 70% most common traces in the first group show the disease progression into a multi-comorbidity state, traces in the second group show patients are often diagnosed with only one category of complications.

3.2 Short-Term Pathways Analysis

In this analysis, we focus on the process of diagnosis for the aforementioned complications in time intervals of 6 months. Cardiovascular and kidney complications are the most frequent in the studied datasets. These categories of diagnoses are the most involved in the progress of diabetes as a process. A total of 10,220 and 1,604 6-month traces, involving 3,217 and 821 patients from the old and young groups, respectively. Process mining and sequencing rules were applied to detect possible pathways that can later be clinically validated. During

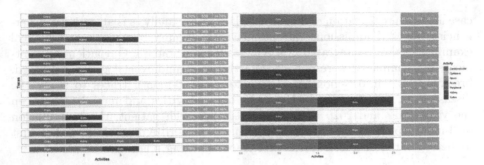

Fig. 1. Top sequences of diabetes complications during long-term diabetes disease progression. These sequences cover 70% of the total number of traces. Left: Traces accounting for patients aged 60 years or older. Right: Patients between 35 and 59 years.

the pro-cess discovery, the most common pathways for diabetes complications in 6-month intervals were extracted. Evaluating the traces, we obtain that the most frequent diagnoses are those related to Chronic Heart Disease (coronary occlusions, hypertensive heart disease...), renal microangiopathies (nephropathy), and brain vascular diagnostics (brain artery occlusions with or without strokes).

Figure 2 shows the most common traces covering 80% of identified traces for both cohorts. This shows differences in both groups. While senior patients may present a diagnosis of heart disease and renal microangiopathy in the short term, younger patients do not often show these complications combined in a 6-month period. The relative frequency of kidney diseases is slightly reduced, while the relative frequency of cerebrovascular affections increases. This show, for example, that in short term, type-2 diabetes senior patients with identified complications suffer from diagnoses related to chronic heart disease in 45% of the 6-month periodical evaluations, while 10% of these 6-month periods are also related to nephropathies and kidney microvascular complications.

Further, the traces can be mapped to identify relations between the most prevalent complications based on different procedural rules. For example, the healthcare process and the recollection of clinical data and diagnoses of both cohorts can be studied, comparing those 6-month traces where cardiovascular and kidney are present with traces in the last 6 months of patients before death. After evaluating cardiovascular and kidney complications are the most prevalent types of diabetes complications, the top common diagnoses can be studied at a lower level.

At a top-level, paths for the diagnosis of cardiovascular and kidney complications can be shown as a process map to study the most significant ones. For example, this method can enable us to identify differences between those 6-month intervals corresponding to the exitus of the patient against standard flows for non-exitus intervals. For this purpose, only traces featuring cardiovascular and kidney diagnoses were considered. Rendering process maps covering the most popular traces for specific diagnoses provides more information about the dif-

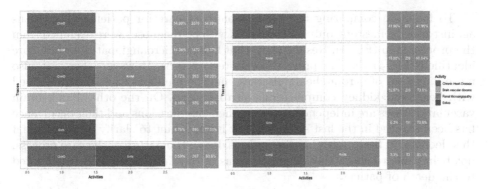

Fig. 2. Top sequences of cardiovascular and kidney complications during traces of 6 months in type-2 diabetes patients. These sequences cover 80% of the total number of traces. Left: Traces accounting for patients aged 60 years or older. Right: Patients between 35 and 59 years.

ferent relations of the top complications in the disease. In this case, pathways for chronic heart disease diagnoses, brain vascular diagnoses, and nephropathies were displayed to cover the top 80% of the most repeated traces. Infrequent traces were removed to reduce noise in the maps. Three major groups can be inferred regarding the most common pathways for the most prevalent cardiovascular and kidney complications. These groups correspond to patients developing brain vascular diseases, those with heart complications, and those with nephropathies, with or without a diagnosed heart disease. Figure 3 presents the registration of diagnoses in the last 6-month period of patients before exitus. It can be noted that half of the patients did not present any diagnosis before death. This result does not mean no diagnosis was identified earlier, but it can be a symptom of patient death occurring when patients are out of hospital surveillance.

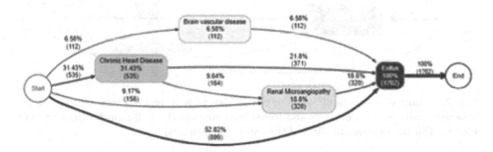

Fig. 3. Cohort of patients 60 years or older. Map of relations between Chronic Heart Disease, Brain vascular diseases and Renal Microangiopathy in the last 6-months activities before Exitus. Top traces covering 80% of the cases were selected.

To facilitate comparing the last 6 months of care for patients with exitus against other semesters, only-exitus traces with no diagnoses were excluded from the process map (Fig. 4). Results show that kidney microangiopathies were more identified (Fig. 4) than in previous semesters (Fig. 5). However, there are also more diagnoses of chronic heart diseases, and the number of patients diagnosed with heart and kidney complications also increases. On the other hand, brain vascular diseases are independent of the other complications and their occurrence has been reduced in the last six months. It is important to clarify in the future if this decrease in brain vascular diagnoses is due to the nature of the care process, not being able to check these complications in the short term, or if it is related to the death of patients.

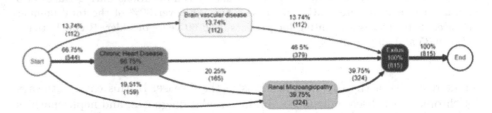

Fig. 4. Cohort of patients 60 years or older. Map of relations between Chronic Heart Disease, Brain vascular diseases and Renal Microangiopathy in the last 6-months activities before Exitus. Top traces covering 80% of the cases were selected. Only-exitus traces with no related diagnoses were masked to facilitate the comparison with Fig. 5.

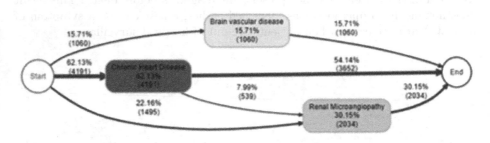

Fig. 5. Cohort of patients 60 years or older. Map of relations between Chronic Heart Disease, Brain vascular diseases and Renal Microangiopathy in 6-month traces without Exitus. Top traces covering 80% of the cases were selected.

Results for younger patients show that these three major complications are rarely related to each other, neither short nor long-term. Therefore, the resulting treatment must have a different scope to those senior patients with muti-comorbidities. In the case of 6-last months for diabetic patients drop-outs, the majority of the cases did not present any diagnoses (Fig. 6). This result shows differences between the two cohorts and diabetes treatments must be designed accordingly. Regarding non-Exitus 6-month traces of young patients, finding the diagnosis of multi-comorbidities is not common at all (Fig. 7). Chronic heart diseases are more common, but the occurrence is slightly lower than in senior patients. By contrast, the proportion of brain vascular diseases increases, also following an independent path.

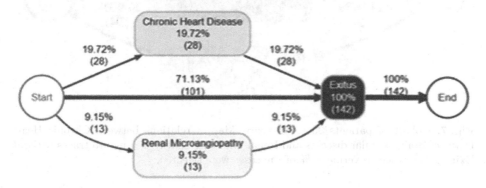

Fig. 6. Cohort of patients 35 to 59 years. Map of relations between Chronic Heart Disease, Brain vascular diseases and Renal Microangiopathy in the last 6-months activities before Exitus. Top traces covering 80% of the cases were selected.

4   Discussion

T2DM prevalence will increase in the next decades and will be proportionally linked to the aging of the population and an increase in life expectancy. Health institutions and healthcare professionals are currently highlighting the importance of reactive care in T2DM. The analysis of complications should be considered to investigate the process of care and therapy and build processes that advance the understanding of patient behavior and individual responses to treatments. Most patients with T2DM have at least one complication, and cardiovascular complications are the leading cause of morbidity and mortality in these patients, as shown by the phenotypes extracted with descriptive statistics. Furthermore, the current literature relates kidney disease and cardiovascular disease in T2DM and international guidelines have been updated to reflect the positive benefits of SGLT2i inhibitors [13]; however, we observe that a significant proportion of patients are in the loop of developing kidney and cardiovascular diseases. The early detection of microvascular complications associated with T2DM is important, as early intervention leads to better outcomes [14]. Nonetheless, this requires awareness of their diagnostic methods, prevalence, and evolution.

In this study, we have implemented process mining tools to understand the relationships of comorbidities throughout the process of care. The application of

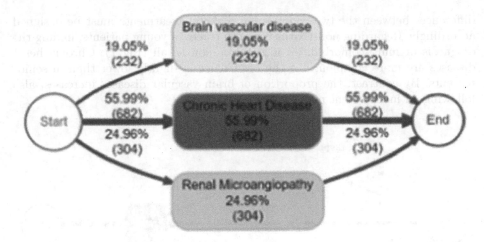

Fig. 7. Cohort of patients 35 to 59 years. Map of relations between Chronic Heart Disease, Brain vascular diseases and Renal Microangiopathy in 6-month traces without Exitus. Top traces covering 80% of the cases were selected.

process mining techniques enabled the identification of clinical pathways which were not identified using descriptive statistics. These outcomes have allowed us to identify common and uncommon paths of patients in the development of cardiovascular disease and other types of microvascular and renal diseases. The information contained in the process model abstractions showed that some patients are diagnosed with a single group of diseases, whereas others go through several types of diagnoses.

Besides the capability of leveraging on longitudinal and temporal insights instead of static data [15], the use of process discovery retrieves process abstractions for the onset of complications highlighted the need to envision the evaluation of a patient condition as an interactive process and not as a phenotype based on clinical variables in the electronic health records. The disease progression flows detected in the specific clinical setting of Hospital La Fe allow for re-assessing risk profiles and characterizing the complexity of T2DM patients. This is especially important in managing a chronic condition such as T2DM, in which we could concentrate information from 8 years of follow-up from routinely collected data in which changes may happen very slowly.

In the context of digital health systems which are embracing new technologies, it is important to find meaningful groups of patients who follow the most frequent patterns of complications, as well as able to identify those critical cases that need specific attention to support clinicians in the complex process of evaluating T2DM patients [16].

Our results show that process mining is useful for gaining a better understanding of type 2 diabetes and other related comorbidities, stratified by age, as it was able to synthesize the variability of the development of complications into clear and understandable patterns, which made it possible to reconstruct clinical pathways [17]. Future work should dig into the process of development of specific cardiovascular and kidney diseases, and the extent to which changes in the treatment can have an impact on the development of complications and glycemic control.

References

1. Munoz-Gama, J., et al.: Process mining for healthcare: characteristics and challenges. J. Biomed. Inform. **127**, 103994 (2022)
2. Buse, J.B., et al.: 2019 update to: management of hyperglycemia in type 2 diabetes, 2018. A consensus report by the American Diabetes Association (ADA) and the European Association for the Study of Diabetes (EASD). Diabetes Care **43**(2), 487–493 (2020)
3. Mauricio, D., Alonso, N., Gratacòs, M.: Chronic diabetes complications: the need to move beyond classical concepts. Trends Endocrinol. Metab. **31**(4), 287–295 (2020)
4. Redondo, M.J., et al.: The clinical consequences of heterogeneity within and between different diabetes types. Diabetologia **63**, 2040–2048 (2020). https://doi.org/10.1007/s00125-020-05211-7
5. Martin, N., et al.: Recommendations for enhancing the usability and understandability of process mining in healthcare. Artif. Intell. Med. **109**, 101962 (2020)
6. De Ramon Fernandez, A., Ruiz Fernandez, D., Sabuco Garcia, Y.: Business process management for optimizing clinical processes: a systematic literature review. Health Inform. J. **26**(2), 1305–1320 (2020)
7. Yang, J., et al.: Brief introduction of medical database and data mining technology in big data era. J. Evid. Based Med. **13**(1), 57–69 (2020)
8. Fernandez-Llatas, C.: Interactive process mining in practice: interactive process indicators. In: Fernandez-Llatas, C. (ed.) Interactive Process Mining in Healthcare, pp. 141–162. Springer, Cham (2021). https://doi.org/10.1007/978-3-030-53993-1_9
9. O'Connell, J.M., Manson, S.M.: Understanding the economic costs of diabetes and prediabetes and what we may learn about reducing the health and economic burden of these conditions. Diabetes Care **42**(9), 1609 (2019)
10. American Diabetes Association: 2. Classification and diagnosis of diabetes: standards of medical care in diabetes-2020. Diabetes Care **43**(Supplement_1), S14–S31 (2020)
11. R Core Team: R: a language and environment for statistical computing. R Foundation for Statistical Computing, Vienna (2018)
12. Janssenswillen, G.: bupaR: business process analysis in R (2023). https://bupar.net/. https://github.com/bupaverse/bupaR/. https://bupaverse.github.io/bupaR/
13. Ryan, P.B., et al.: Comparative effectiveness of canagliflozin, SGLT2 inhibitors and non-SGLT2 inhibitors on the risk of hospitalization for heart failure and amputation in patients with type 2 diabetes mellitus: a real-world meta-analysis of 4 observational databases (OBSERVE-4D). Diabetes Obes. Metab. **20**(11), 2585–2597 (2018)
14. Papatheodorou, K., Banach, M., Bekiari, E., Rizzo, M., Edmonds, M., et al.: Complications of diabetes 2017 (2018)
15. Dagliati, A., et al.: Inferring temporal phenotypes with topological data analysis and pseudo time-series. In: Riaño, D., Wilk, S., ten Teije, A. (eds.) AIME 2019. LNCS, vol. 11526, pp. 399–409. Springer, Cham (2019). https://doi.org/10.1007/978-3-030-21642-9_50

16. Contreras, I., Vehi, J.: Artificial intelligence for diabetes management and decision support: literature review. J. Med. Internet Res. **20**(5), e10775 (2018)
17. Dagliati, A., Tibollo, V., Cogni, G., Chiovato, L., Bellazzi, R., Sacchi, L.: Careflow mining techniques to explore type 2 diabetes evolution. J. Diabetes Sci. Technol. **12**(2), 251–259 (2018)

From Script to Application. A bupaR Integration into PMApp for Interactive Process Mining Research

Roberto Tornero-Costa[1]([✉]) (iD), Carlos Fernandez-Llatas[1,3] (iD), Niels Martin[2] (iD),
Gert Janssenswillen[2] (iD), and Gerhardus A. W. M. van Hulzen[2] (iD)

[1] Universitat Politècnica de València, 46022 Valencia, Spain
{rotorcos,carferll}@itaca.upv.es
[2] Hasselt University, 3500 Hasselt, Belgium
{niels.martin,gert.janssenswillen,gerard.vanhulzen}@uhasselt.be
[3] CLINTEC Karolinska Institutet, Stockholm, Sweden

Abstract. There are several open-source Process Mining tools available for research purposes. PMApp is an Interactive Process Mining toolkit developed to facilitate process discovery and analysis in healthcare. PMApp is designed to introduce healthcare professionals to Process Mining, simplifying the learning process by reducing the need for extensive coding knowledge. With PMApp, users can easily apply process mining techniques to healthcare data without the steep learning curve typically associated with coding languages. At the same time, bupaR is an open-source, integrated suite of R-packages for handling and analysing process data. bupaR provides support for different stages in process analysis and offers the flexibility of coding tools. Given the complementarity between both tools, this paper outlines how PMApp has been extended to integrate R code and bupaR functionalities. This integration provides an opportunity to extend research in process mining by combining two powerful tools. The paper will showcase the feasibility of this integration and demonstrate how researchers and clinicians can leverage the combined capabilities of Interactive Process Mining and scripting open-source tools to gain insights into complex healthcare processes.

Keywords: Interactive Process Mining in Healthcare · Interfaces for Process Oriented Data Science · Process Mining · R · Interactive Process Mining

1 Introduction

Interactive models in healthcare decision-making facilitate human-machine interaction, placing human expert knowledge at the heart of the system. While computers provide the best memory and mathematical capabilities, human experts can coordinate computation towards the truth models [1]. However, the application of Interactive Methodology requires the involvement of human experts

© The Author(s), under exclusive license to Springer Nature Switzerland AG 2024
J. M. Juarez et al. (Eds.): XAI-Healthcare/PM4H 2023, CCIS 2020, pp. 107–117, 2024.
https://doi.org/10.1007/978-3-031-54303-6_11

in the process of learning. This requires acceptance from professionals. Interactive Process Mining (IPM) toolkits adapt these methodologies in a human-understandable way, facilitating human-system interaction [2]. As stated in the 2022 PODS4H report for process mining for healthcare, process mining is a research field that involves a multidisciplinary team, with rapid evolution and constant advances in novel paradigms [3]. Incorporating IPM tools is crucial for engaging healthcare professionals in a self-directed process mining analysis. To effectively integrate process mining applications and their results into clinical practice, it is vital to encourage non-technical healthcare stakeholders to participate. For this purpose, more easily human-understandable learning systems and interactive methodologies are required. However, developing these technologies often lacks the flexibility of coding scripts which facilitate quickly implementing newly developed algorithms and methods into research using open-source libraries. At the same time, coding scripts may introduce new barriers to the adoption of process mining methodologies by healthcare professionals. The goal of this paper is to present an integration of scripting languages in an interactive process mining toolkit, easing the adoption of new open-source algorithms. The R language and the open-source bupaR suite have been integrated into the PMApp IPM software to boost the flexibility of implementing new paradigms and increasing the flexibility of the toolkit. The bupaR functionalities have been integrated into the experiment designer, allowing the users to create R pipelines to manipulate datasets.

2   Background

PMApp is an Interactive Process Mining (IPM) software developed in C# with .NET core, and composed of several tools that ease the analysis of processes with some user-friendly interfaces. This software is based on the PALIA Suite tool [4]. Since then, this tool has been successfully adapted and utilized in various healthcare projects [5,6]. Starting in 2020, these tools were integrated into a new IPM software, PMApp. This application is designed to facilitate non-code interactive analysis. The experiment designer interface enables the processing of datasets to define an activity log. A collection of block-shaped elements supports non-code users to implement their own processing tools and to manipulate the log. In addition, the application enables users to define standardizations, create processes, and apply different models such us Timed Parallel Automatons (TPA) [7]. Once the activity log is created and the experiment is designed, end-users can perform their custom statistical and visual analysis, create reports, and save the data without any coding knowledge required. This software allows non-technical stakeholders, such as healthcare clinicians, to be involved in the analysis. Process mining experts can design experiment pipelines to move the research process into the clinical environment through iterations with clinician experts. Next, process mining and code experts can save the experiment design to make it easier for clinicians to reuse the models without inputting code or modifying the pipeline. Figure 1 introduces the two major interfaces of the application, and Fig. 2 presents the block-shaped element supporting the integration of R language into the experiment designer of the application.

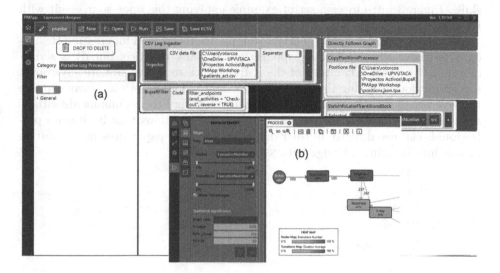

Fig. 1. Presentation of (a) the interface to design the IPM experiment - definition of the activity log, manipulation through filters, definition of the processes, etc. -, (b) the interface of the IMP analysis, where the user is presented with a map of the process, different options to change perspectives, implement statistical analysis, etc. In this figure, both interfaces are presented overlapped.

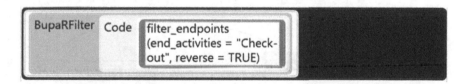

Fig. 2. Example of block element embedding of R language into the PMApp experiment designer.

bupaR, also known as bupaverse, is an open-source, integrated suite of R-packages for the handling and analysis of business process data [8], which provides extensive support for conducting process analysis using scripts in R. The suite is organized around the main package bupaR. Together with other packages, each supporting specific stages of process analysis, such as descriptive statistics (edeaR), visualizations (processmapR), and conformance checking (process-checkR), it forms the core of the ecosystem. Beyond this core, it is complemented with extension packages for other use cases, such as process prediction, event log construction, trace clustering, etc. Due to R's usefulness in quickly implementing code scripts, bupaR allows users to conduct fast research on process mining. However, it requires users to have coding knowledge and some programming skills. Figure 3 introduces a script example applying this open-source suit with RStudio IDE.

While PMApp is a Desktop oriented application designed for creating Health-experts-usable Process Mining applications in daily practice, bupaR is script-oriented that takes advantage of the R-framework to create statistically robust algorithms for analyzing the process. The combination of both tools can suppose the leverage of the power of bupaR for easily creating Process mining algorithms with the capabilities of PMApp to develop final solutions that healthcare professionals can use directly in their daily routine. This paper presents a solution in this line, creating a bridge between both worlds.

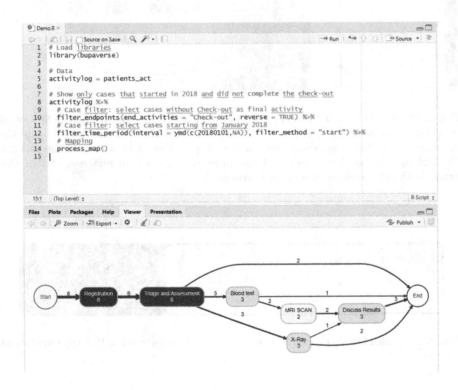

Fig. 3. Example of a process analysis by R using an integrated development environment (IDE). In this case, the image shows the Rstudio IDE v 2023.3.0.386 [9] for R scripts.

3   Use Case and Test Scenario

The bupaR ecosystem provides both real-life and artificial event logs. For demonstration purposes, we will use the 'Patients' activity log. This synthetic data comprises 500 emergency admissions to a hospital between 2nd January 2017 and 5th May 2018. The event log is composed of 7 different activities, from the 'Registration' of the patient to the 'Check-out'. Between these initial and final activities, the process follows with 'Triage and assessment', some medical examinations ('Blood test', 'MRI scan' and 'X-Ray'), and a final 'Discussion of results' that should lead to the final activity and end of the process. This activity log provides a generalized and clear version of an emergency process [10] and facilitates a clean test to evaluate the integration of bupaR and PMApp.

The integration of R engine and bupaR ecosystem into PMApp was implemented through R.NET library [11] facilitating the compatibility of .NET framework and R. Through this integration, any set of R function and parameters can be implemented as part of the process mining experiment pipeline in PMApp. For demonstration purposes, a set of tests were designed to evaluate the proper integration of R, applying bupaR to manipulate the log and integrate the results in the PMApp software.

The synthetic process in the activity log is shown in Fig. 4. The activity log comprises two major pathways with minimal noise. Also, it can be noted immediately that not every trace completes the final "Check-out" activity. The following cases were evaluated in PMApp environment and implemented in R (bupaR) as a control. The table below summarizes the implementation of multiple functions with direct R code into the IPM tool. The control was implemented in R v4.2.2 [12] with bupaR v.0.5.2 library [13]. The PMApp integration of R applies to the same version and libraries installed in the R path. Table 1 summarises the 3 tests that are implemented to evaluate the integration of bupaR into PMApp application.

Table 1. Test cases designed to evaluate the integration of R and bupaR into an IPM application (PMApp).

Test case	Set of bupaR functions	Role in the experiment designer
a. Filtering cases based on activities and timestamps	filter_activity_presence(), filter_endpoints(), filter_time_period()	Integrate open-source script methods to select activities of cases that fulfill different requirements
b. Data augmentation and conditional filtering	throughput_time(), augment(), filter()	Integrate open-source script methods to compute new trace or activity variables, incorporate them into the log and filter activities based on them
c. Manipulation of activities	act_unite()	Integrate open-source script methods to manipulate the definition of activities into the process, i.e. merging different activities into one

(a)

(b)

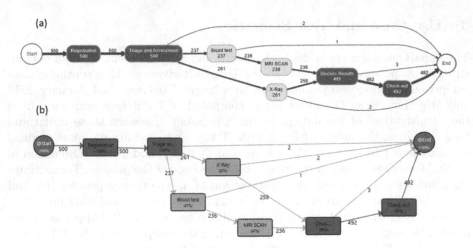

Fig. 4. Process map of'patients' event log in (a) R (bupaR) and (b) PMApp software.

3.1   Filtering Cases Based on Activities and Timestamps

In order to detect incomplete cases, we could implement an R script defining which cases of the activity we want to filter. In this case, the bupaR library allows the user to select those cases whose last activity is "Check-out" by the filter_endpoints() function. Selecting the parameter "reverse = TRUE", we inverse the selection, only keeping incomplete cases. We could also apply filter_time_period() with parameter filter_method = "start" to select only cases that began in 2018. Using the process_map() function, we can render a static map of the filtered cases. If the user was interested in a deeper analysis of these cases, coding more lines and calling other functions would be required.

```
library(bupaverse)
patients_act %>%
   # Case filter based on activity
   filter_endpoints(end_activities = "Check-out",
                    reverse=TRUE) %>%
   # Case filter based on dates
   filter_time_period(
              interval = ymd(c(20180101,20180630)),
          filter_method = "start"
       ) %>%
process_map()
```

Once the log has been defined, the PMApp software enables a set of interactive tools to implement deeper analyses. However, designing different methods to manipulate the log requires declaring new C# functions in the program. The integration of a scripting language into the software enables implementing new methods in an easy way, even working with the compiled application. Also, previous open-source R libraries and functions can be called. Therefore, users do not need to script already implemented algorithms in open-source libraries. Figure 5 shows the creation of the activity log in the PMApp application and the integration of the R script as a block in the experiment design. Once the experiment has been designed, it can be saved to facilitate reusing. The resulting activity logs are equal, and the process maps rendered in the R script and the PMApp application are able to show the same information, as can be seen in Fig. 6. Later, the PMApp application allows the end-user to modify the map with different statistical features and other tools without the need to integrate more code.

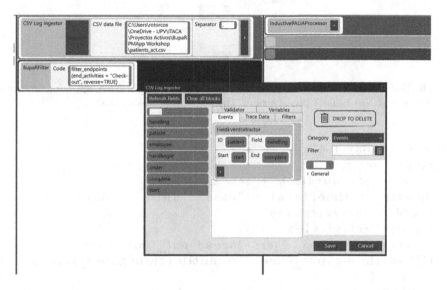

Fig. 5. Creation of activity log and implementation of bupaR function in the PMApp experiment designer. The bupaR filter block allows to implement R code to modify the activity log.

3.2   Data Augmentation and Conditional Filtering

The integration of an R script also allows to compute new trace or activity metadata to expand the activity log without defining new C# algorithms. Later, these new variables can be applied to manipulate the log, e.g. applying filters, or integrated into the PMApp experiment for further analysis. For example, users could define a pipeline to ingest only an activity log into the analysis that comprises longer-duration cases. For demonstration purposes, we will only select cases where the total throughput time is in the third quantile (Q3) or higher. To create a quick process map in an R script, we can execute:

(a)

(b)

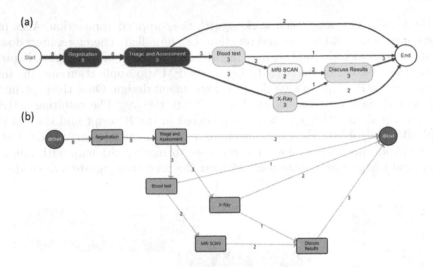

Fig. 6. Test 1. Selection of cases without Check-out activity in 2018, applying developed functions in R. (a) bupaR process map. (b) PMApp process map.

```
library(bupaverse)
patients_act %>%
  # calculate duration trace per case
  throughput_time(level = "case", units = "days") %>%
  # add to activity log
  augment(activitylog) %>%
  # keep only cases where throughput_time > q3
  filter(throughput_time > quantile(throughput_time,0.75)
    ) %>%
  # Add to the process map new metrics
  process_map(
    type_nodes = frequency("relative_case"),
    type_edges = performance(units = "hours"),
    sec_nodes = frequency("absolute"),
    sec_edges = frequency("absolute")
  )
```

This code can be integrated to manipulate the defined activity log in the PMApp experiment designer. Then, the 'Enhancement' PMApp tool allows the end-users to modify the resulting map interactively. Later, other analyses can be conducted without requiring adding more code. Figure 7 shows the R and PMApp outputs.

(a)

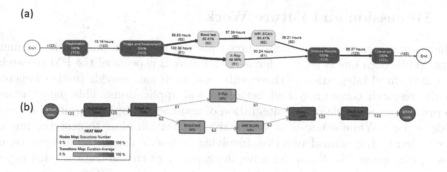

(b)

Fig. 7. Test 2. Integration of bupaR functions to map the activities of only those patients with the highest throughput time (quantile 3). (a) bupaR process map. (b) PMApp process map.

3.3 Manipulation of Activities

Although PMApp already supports categorization, grouping and renaming activities in the log. A user might want to implement a new definition for some activities in an integrated R script. For example, in the 'patients' log, there are two pathways which comprise two different medical image techniques, MRI scan and X-Ray image. These techniques could unite into a higher category: "Medical image". This is easy to implement using only one function, without the need to add more PMApp blocks into the experiment designer. Figure 8 display the results for this test.

```
library(bupaverse)
patients_act %>%
    act_unite('Medical image' = c("MRI_SCAN","X-Ray"))
        %>%
    process_map()
```

(a)

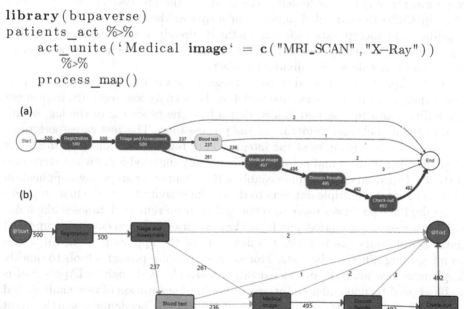

(b)

Fig. 8. Test 3. Modifying the log by renaming different activities with one function. (a) bupaR process map. (b) PMApp process map.

4   Discussion and Future Work

There are currently several open-source suites for process mining in the community. While each one is slightly focused on different aspects of the PM research, a standardized integration of these suites and tools can provide further benefits to the research community and better clinical applications. This paper aimed to provide an example of the integration of non-code compiled applications and code scripts. While a non-code interactive application aims to ease the implementation in the clinical practice, involving non-code users, the integration of a script language like R can improve the capacity of customization. This paper demonstrates the potential of better cooperation between different open-source IPM and PM suites.

For research purposes, the development of integration methods between the different suites reduces the learning curve to integrate newly developed functionalities and algorithms into the research workflow of every academic group. In addition, development and implementation times for existing algorithms are reduced, allowing more time to be allocated to pure research. Moreover, the combination of tools developed by different research groups may allow new collaboration opportunities.

For clinical practice, PMApp's vision is to provide a framework for non-technical users to carry out different studies with defined processes and logs in the experiment designer. While an interactive interface can facilitate this transition for clinicians, the creation of the pipeline which will set the experiment in the experiment designer is performed by a technical profile which can implement the requirements of clinicians to define the needs for the process. The incorporation of scripts into the compiled application increases the flexibility to create new pipelines and incorporate different methods already implemented in different PM libraries, reducing the burden for constant development of new versions based on requirements of individual projects.

In PMApp, there are fixed steps to design a log and the analysis in the experiment designer. Firstly, users have to define the activity log; secondly, implement some filters into the log and finally, define how the processor of the log, definition of the models and rendering of the process map. The test cases performed in this paper are examples of the integration of bupaR functions into the filters step, allowing to filter and manipulate an already ingested log. Future steps into further integration could include enabling R code in other steps, e.g., application of R to integrate multiple datasets to define the activity log at the first step. For example, PMApp allows users to retrieve data from remote databases like MongoDB to create automated pipelines that end-users can execute with updated data. Script codes can improve the flexibility of the application when it comes to integrating different datasets. Moreover, scripts are powerful tools to quickly implement new analysis and visualization methods. Although PMApp's goal is to be applied by non-coder end-users, future implementation of new analysis and visualization techniques could create new synergies in academic research. Right now, only the bupaR library has been integrated into the system as a demo. However, other libraries in the bupaverse ecosystem or other open-source could facilitate the creation of better pipelines, for example the application of declarative rules for conformance checking in a few lines with the library processcheckR or PMineR [14].

References

1. Fernández-Llatas, C., Meneu, T., Traver, V., Benedi, J.-M.: Applying evidence-based medicine in telehealth: an interactive pattern recognition approximation. Int. J. Environ. Res. Public Health **10**(11), 5671–5682 (2013)
2. Fernandez-Llatas, C.: Interactive Process Mining in Healthcare. Springer, Cham (2021). https://doi.org/10.1007/978-3-030-53993-1
3. Munoz-Gama, J., et al.: Process mining for healthcare: characteristics and challenges. J. Biomed. Inform. **127**, 103994 (2022)
4. Fernandez-Llatas, C., Lizondo, A., Monton, E., Benedi, J.-M., Traver, V.: Process mining methodology for health process tracking using real-time indoor location systems. Sensors **15**(12), 29821–29840 (2015)
5. Fernandez-Llatas, C., et al.: Analyzing medical emergency processes with process mining: the stroke case. In: Daniel, F., Sheng, Q., Motahari, H. (eds.) BPM 2018. LNBIP, vol. 342, pp. 214–225. Springer, Cham (2019). https://doi.org/10.1007/978-3-030-11641-5_17
6. Ibanez-Sanchez, G., et al.: Toward value-based healthcare through interactive process mining in emergency rooms: the stroke case. Int. J. Environ. Res. Public Health **16**(10), 1783 (2019)
7. Fernandez-Llatas, C., Pileggi, S.F., Traver, V., Benedi, J.M.: Timed parallel automaton: a mathematical tool for defining highly expressive formal workflows. In: 2011 Fifth Asia Modelling Symposium, pp. 56–61. IEEE (2011)
8. Janssenswillen, G., Depaire, B., Swennen, M., Jans, M., Vanhoof, K.: bupaR: enabling reproducible business process analysis. Knowl.-Based Syst. **163**, 927–930 (2019)
9. Allaire, J.: RStudio: integrated development environment for R, Boston, MA **770**(394), 165–171 (2012)
10. Martin, N., Swennen, M., Depaire, B., Jans, M., Caris, A., Vanhoof, K.: Retrieving batch organisation of work insights from event logs. Decis. Support Syst. **100**, 119–128 (2017)
11. Perraud, J.-M.: R.net (2017)
12. R Core Team: R: a language and environment for statistical computing. R Foundation for Statistical Computing, Vienna (2018)
13. Janssenswillen, G.: bupaR: business process analysis in R (2023). https://bupar.net/. https://github.com/bupaverse/bupaR/. https://bupaverse.github.io/bupaR/
14. Gatta, R., et al.: pMineR: an innovative R library for performing process mining in medicine. In: ten Teije, A., Popow, C., Holmes, J., Sacchi, L. (eds.) AIME 2017. LNCS, vol. 10259, pp. 351–355. Springer, Cham (2017). https://doi.org/10.1007/978-3-319-59758-4_42

Understanding Prostate Cancer Care Process Using Process Mining: A Case Study

Zoe Valero-Ramon[1]([✉])(iD), Carlos Fernandez-Llatas[1,3](iD), Gonzalo Collantes[2](iD), Bernardo Valdivieso[2](iD), and Vicente Traver[1](iD)

[1] Universitat Politècnica de València, Camino de Vera s/n, Valencia, Spain
{zoevara,cfllatas,vtraver}@itaca.upv.es
[2] Hospital La Fe, Valencia, Spain
gonzalo_collantes@iislafe.es, valdivieso_ber@gva.es
[3] Karolinska Institutet, Stockholm, Sweden
carlos.fernandezllatas@ki.se

Abstract. Prostate cancer is the fourth most common cancer in the EU-27, with around 470,000 new cases yearly, and the most common cancer among males. Patients diagnosed with prostate cancer go through established procedures, and the decisions made about the treatments are vital due to cancer's unfavorable essence evolution. In this context, prostate-specific antigen tests are helpful in stratifying surveillance and subsequent risk and are monitored for relapsed detection after diagnosis and during and after treatment. Electronic Health Records store longitudinal data and record detailed cancer therapies and PSA values during this process. Incorporating this information and the temporal perspective into the risk models could stratify patients with similar evolution. The perception of clinical processes behind treatments. Applying Process Mining techniques and an interactive paradigm with the Dynamic Risk Models framework could result in the definition of new PSA evolution groups, enabling prostate cancer experts to control disease favorably and most appropriate treatments. This work uses real-world data from prostate cancer patients collected in a public hospital and Process Mining techniques to obtain new behavioral models for PSA evolution. The results represent prostate cancer care processes for different PSA evolution groups, allowing looking for awareness and differences.

Keywords: Process Mining · Prostate cancer · Patient's progress · Care process

1 Introduction

Prostate cancer is the fourth most common cancer in the EU-27, with around 470,000 new cases a year, and the most common cancer among males (22.2% of the male total) [2]. Although it often has an indolent course, prostate cancer remains the third-leading cause of cancer death in men.

© The Author(s), under exclusive license to Springer Nature Switzerland AG 2024
J. M. Juarez et al. (Eds.): XAI-Healthcare/PM4H 2023, CCIS 2020, pp. 118–130, 2024.
https://doi.org/10.1007/978-3-031-54303-6_12

Consistently, prostate cancer suspicious due to elevated Prostate-specific antigen (PSA) or pathological digital rectal examination is confirmed by a transrectal or transperineal ultrasound-guided biopsy, following the recommendations of the international guidelines [19]. Then, a PSA test is valuable to stratify surveillance and subsequent risk [6]. According to the disease risk, and based on the Gleason grade, the PSA value, and the tumor extension, the health experts consider the most appropriate treatment options for the patient. Many prostate cancers are slow-growing, requiring long-term follow-up in PSA screening. Notwithstanding, PSA is monitored for relapse detection after a diagnosis and treatment.

In this context, a large quantity of data is collected during the delivery of cancer treatment to witness the care received by the patient, specifically regarding the PSA tests. Patients diagnosed with cancer go through detailed therapies, and PSA tests are performed within a periodicity. Electronic Health Records (EHR) are widely used to store longitudinal data and record patients' information, such as vital signs, medications, or laboratory values, including PSA test values and information about diagnoses and treatments. Nowadays, multiple management options exist for men diagnosed with prostate cancer, and advances in the diagnosis and treatment of prostate cancer have improved the ability to stratify patients by risk and allowed clinicians to recommend therapy [11]. Incorporating the EHR's existent information and the temporal perspective into the risk models makes it possible to stratify patients with similar evolution and enhance the perception of clinical processes behind treatments. Understanding the cancer cycles and care process is crucial for creating better and more effective treatments. Analyzing these processes using data-driven techniques can offer a new perspective on cancer perception. One of these promising techniques is Process Mining [18]. Process Mining enables experts to understand the process of creating human-readable processes that explain the actual processes based on the automatic learning of event data available on databases. Process Mining has been successfully used in lots of cases [14]. Exploring the event logs associated with EHR, related to cancer treatment and PSA values using Process Mining is a promising way to support the comprehension and improve the quality of cancer care processes [7,10,12].

In this regard, Process Mining can construct individual and human behavior models [5] to stratify patients. From now on, [17] proposed a dynamic approach for risk models based on the stratification groups that permit a better comprehension of the clinical cases. This work defined the Dynamic Risk Model as *the behavioral categorization of a disease considering the evolution of the associated risk models from a dynamic perspective, permitting a better understanding of the patients' groups' progression*. It could allow us to include the variability and evolution of concrete variables for prostate cancer. In this case, the PSA was used to obtain groups representing different behaviors and gain insights into the prostate cancer processes. Relapse detection requires PSA to reach a minimum and increase, which may result in delays in applying further needed treatments.

The definition of PSA evolution groups would enable health experts to control diseases favorably and select the most appropriate treatments.

The experts' involvement is essential when considering the dynamic risk models and the evolution of the associated variables. Health professionals should be able to pose relevant and clinical-related questions about the process and incorporate their knowledge by correcting, verifying, and refining the results. It implies the interaction between the health professionals and the techniques and results [20]. The Interactive paradigm defines this concept by integrating human activity into the process [4]. It guarantees convenient cooperation between the human expert and the techniques to produce understandable results for the actual process and allows improvement according to human knowledge.

This work uses real-world data from prostate cancer patients collected in a public hospital and Process Mining techniques to obtain new behavioral models for PSA evolution. Applying the Interactive paradigm through implementing the Dynamic Risk Models framework resulted in the patients' stratification based on their PSA evolution, looking for awareness and differences in prostate cancer care processes. Furthermore, the work intends to advance in the PM4H community by contributing to the specific challenges proposed by [13]. Concretely, we consider the work presented relevant for challenge C1 - Design Dedicated/Tailored Methodologies and Frameworks -and challenge C4 - Deal with Reality. On the one hand, the paper puts into practice an interactive and question-based methodology for deploying Dynamic Risk Models using Process Mining that guides Process Mining and health experts through the different stages of the analysis. On the other side, the work uses real data from a hospital system and the active involvement of its health experts to obtain and validate the results, as stated in challenge 4. The organization of the remainder of this article is as follows. After the introduction, a brief description of the methods and tools employed in Sect. 2 is included. Then, Sect. 3 presents the experimental results. Finally, discussion and conclusions are drawn in Sect. 4.

2   Materials and Methods

The work presented in [15] proposes an *interactive and question-based methodology* [14] *for deploying Dynamic Risk Models for chronic conditions using Process Mining*. This formal methodology incorporates clinical experts' needs following the Interactive Process methodology [4] to provide understandable results, considering the nature of the chronic disease during the formulation. The methodology was designed and tested to analyze chronic disease's underlying processes but could be used to develop Dynamic Risk Models focused on other diseases or pathologies. In this case, we aimed to apply the framework for prostate cancer to validate the viability of its results in this field.

Overall, the methodology is conceived to support clinical and Process Mining experts –process miners– in understanding chronic disease's progress. It gives them a straightforward guideline for applying Interactive Process Mining to analyze the disease's underlying processes. The clinical experts can use the methodology to commit to their condition deeply by suggesting the appropriate questions and validating and correcting the results. On the contrary, the Process Mining professionals support their understanding of Process Mining techniques. It introduces six steps as follows (Fig. 1): (1) defining the corresponding question; (2) analyzing the risk factors and variables that could be used for answering the posed question; (3) verifying data quality and availability; (4) formalization of the Interactive Process Indicator (IPI) associated with the posed question; (5) applying stratification analysis; and (6) validating the results with the experts in the field.

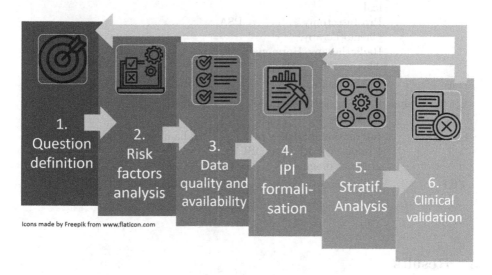

Fig. 1. Interactive and question-based methodology for deploying Dynamic Risk Models for chronic conditions using Process Mining. Presented in [15]

We used the LifeChamps version of PMApp [16] as a Process Mining tool within the methodology. PMApp for LifeChamps is an exploratory analytical Interactive Process Mining tool with clinical relevance for healthcare professionals monitoring cancer patients' care processes. It allows the visualization of real-world cancer process data from health organizations and self-reported by cancer patients. The tool produces advanced process views to empower the analysis made by the health stakeholders [8].

In this work, we collaborated with the Hospital La Fe in Valencia, a publicly owned and managed hospital responsible for the health care of 300,000 inhabitants. The data set included 1,267 unique patients with a prostate cancer diagnosis. The data included seven variables with information at diagnosis and related to cancer care. They were retrospectively collected from the hospital IT systems from May 2011 to April 2022 (see Table 1 for the complete data set). Moreover, it should be mentioned that ethical approval was obtained for this study from the Ethical Committee of Hospital Universitario La Fe on 31st

March 2021 (Registration number: 2019-157-1). Before their transfer, all samples were entirely anonymized by the hospital's IT department.

Table 1. Data set.

Column Name	Description	Type	Example
Patient ID	Global unique identifier	Numeric	24972
Age	Patient age at diagnose	Numeric	
Episode	Episode name: Diagnose, Radical Prostatectomy, PSA, Chemotherapy, Radiotherapy, Hormonal Therapy, or Exitus	Alphanumeric	Chemotherapy
Date	Episode date	Date	18MAY2021
Episode Value	PSA (ng/mL) value only for PSA episode	Numeric	20.94
Grade	International Society of Urological Pathology (ISUP) grade at diagnosis	Alphanumeric	Grade group(1–5)
TR1	Initial group treatment	Alphanumeric	ADT

3   Results

As stated, the work had three main goals. On the one hand, to obtain prostate cancer processes using real-world data and Process Mining techniques. On the other hand, to characterize patients using PSA evolution through applying the Dynamic Risk Models framework. Lastly, to validate the cancer field's approach using the framework.

For this purpose, we applied the methodology introduced in Sect. 2. Within this methodology, the prostate cancer care expert's involvement is crucial throughout all steps, from the definition of the question to the formalization of the IPI and its clinical validation. With this in mind, we designed a set of interactions between the two world experts, cancer prostate and process miner. Concretely, we implemented three collaborative working sessions, which details are included in Table 2. After each joint session, process miners worked with the data using the PMApp LifeChamps version to implement an IPI preliminary version to be analyzed in the next working session until the IPI evolved to its final version.

During the first session, we focused on understanding the clinical problem and the data by the process miners and the definition of the questions by the health experts. For this, we obtained a preliminary IPI presenting the care process

Table 2. Working sessions detail

Working session 1	
Objectives	To understand the data set and the clinical problem To define the question and the IPI
Experts involved	Biomedical engineering Process miners
Framework steps covered	1. Question definition 2. Risk factors analysis 3. Data quality and availability
Session 2	
Objectives	To validate the preliminary IPI To perform data curation To analyse PSA values and its incorporation to the IPI To evaluate differences among groups of patients
Experts involvement	Urologist and biomedical engineering Process miners
Framework steps covered	4. IPI formalization 5. Clinical validation 1. Redefinition of the question
Session 3	
Objectives	Validation of the new IPI version Evaluation of the different PSA groups
Experts involvement	Health experts: biomedical engineering Process miners
Framework steps covered	5. Stratification analysis 6. Clinical validation

with the different clinical events (diagnosis, radical prostatectomy, chemotherapy, radiotherapy, hormonal therapy, and *exitus*) presented by the nodes. The arrows represent the transitions among them. Finally, we added the rest of the information as aggregated data. The result is shown in Fig. 2. As explained in Sect. 2, Process Mining understands data as recorded event logs, where each event refers to a case, an activity, and a point of time to discover, monitor, and improve actual processes. In healthcare, each *Event* contains timestamp information about a patient's healthcare episode. A set of events corresponding to the same patient is called *Case*, and a *Log* is a set of cases. Working with the prostate cancer expert, the PSA level was considered a risk factor during the second step. It is a biochemical measure of the disease's response, progression, and, eventually, recurrence.

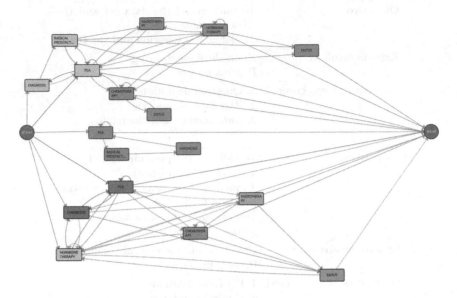

Fig. 2. Prostate cancer process at Hospital Universitario La Fe. The redder color in nodes, as opposed to green, represents a larger number of patients in the stage, while the redder color in transition means a larger number of cases in that transition. (Color figure online)

This information was the input for the process miners' experts to work with the data and formalize the first version of the IPI. During this step, we performed some data curation. We detected and corrected the information for patients with treatment after *exitus* and treatments before diagnosis (but with the same date). In the first case, all information after *exitus* was not incorporated into the later analysis. Concretely, we deleted information from 13 patients.

During the second session, the main objectives were to validate the data curation and the best approach for incorporating the PSA evolution into the IPI. We performed an interactive data curation with the experts, navigating within the process and highlighting the problems with the data. Then, we completed the corresponding corrections following the health experts' knowledge. We corrected the traces, making the diagnosis the first step. On the other hand, we considered the PSA evolution obtaining its process to discretize the PSA value (included in the episode value, see Table 1) with two cut-off points –4 and 10– according to the National Cancer Institute guidelines of the United States [9].

Following the procedure and the methodology, we used the PSA level and the specified cut-off points for grouping patients into different behaviors regarding the PSA evolution. We used trace clustering techniques to group traces with similar behavior, maximizing differences with the rest of the groups to determine patterns. Thus, using the LifeChamps version of PMApp, the PALIA Discovery algorithm, and trace clustering techniques were applied as stratification filters to extract the sub-logs from the main log, representing sub-populations depending on the PSA evolution. We used the Quality Threshold Clustering (QTC) [1] with a Heuristic distance. The QTC algorithm requires a quality threshold to decide the maximum distance among traces in the cluster, where a minimum distance (0.0) means equal flows and a maximum distance (1.0) produces traces with no common activities. Several threshold distances were used to obtain the most significant groups of patients, balancing the number of groups and the behavior shown within each group and looking for clear patterns regarding the PSA evolution. The most meaningful results were obtained for a similarity percentage of 10% and a threshold of 0.2 with seven groups. It resulted in 18 groups (see Figs. 3, 4 and 5).

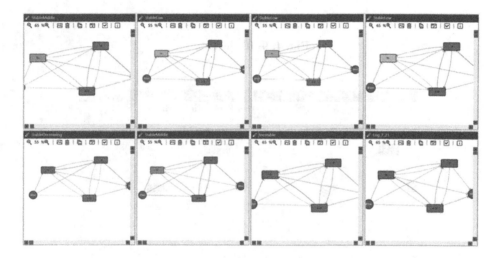

Fig. 3. 1–8 clusters.

The groups were presented to the clinical experts following the methodology. Clinical experts merged them into five different behaviors plus an outliers group and proposed semantic names based on the behavior and characteristics of the PSA evolution (see Table 3). Group 1 –*Stable Low* PSA– has 537 patients. Group 2 represents the *Stable Middle* evolution with 263 patients. Group 3 shows the study's 217 outliers. Group 4 includes 109 patients with a *Stable Decreasing* PSA evolution. Group 5 represents the *Stable High* evolution with 81 patients. Finally, group 6 presents the *Unstable* patients with 60 patients.

These dynamic PSA groups were added as a new variable to the current process. Therefore, we had a new variable to characterize the population, the PSA clusters. Using this new variable, we obtained the care process for each group and looked for new insights working with the clinical experts. Figures 6 and

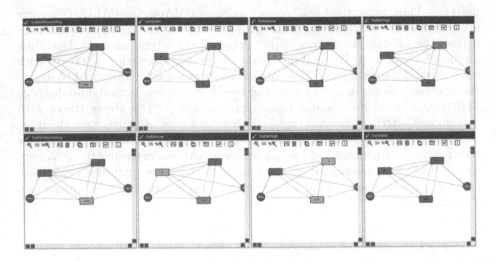

Fig. 4. 9–16 clusters.

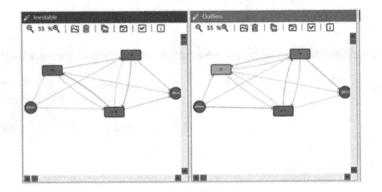

Fig. 5. 17–18 clusters.

Table 3. PSA behavioral groups.

Group	Behavior	Population	%
1	*Stable Low*	537	42.4
2	*Stable Middle*	263	20.8
3	*Outliers*	217	17.1
4	*Stable Decreasing*	109	8.6
5	*Stable High*	81	6.4
5	*Unstable*	60	4.7

Fig. 7 represent two of the six care processes for the *Stable Low PSA evolution* and *Stable High PSA evolution* clusters, respectively.

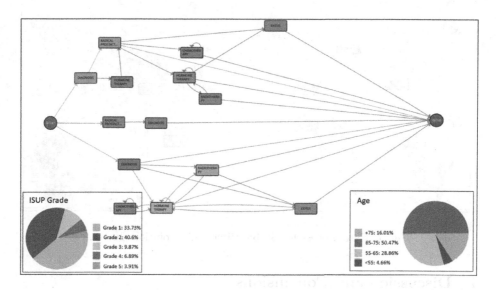

Fig. 6. Care process for Stable Low PSA evolution cluster.

By analyzing these two groups of patients and their care processes, it can be noticed that differences are perceived. The first group, Low PSA, includes three *circuits* representing the three main treatments: radical prostatectomy, radical prostatectomy combined with other treatments (hormonal therapy, radiotherapy, and chemotherapy), and treatment without radical prostatectomy. Within this group, the *exitus* rate, 9%, is lower than the other group. In the second presented group, High PSA, there are two *circuits*, patients treated without radical prostatectomy and radical prostatectomy with or without hormonal therapy. Thus, a difference can be detected by analyzing the groups in the care process. Comparing the *exitus* rate, in this case, it is 26%, higher than in the other case. There are also differences in the population distribution between the two clusters regarding the age and ISUP grade at diagnosis (see pies graphics in both, Fig. 6 and 7). The ISUP grades 1–5 system has the dual benefit of predicting patient outcomes and facilitating patient communication [3]. In the Low PSA, most of the population has better-predicted outcomes based on the ISUP, as 78% were grade 1 or 2, and only 3.9% were grade 5 at the diagnosis. In the High PSA, grades 1 and 2 represent half the population, 54%, and grade 5 is the 23%. We also added the age distribution to the clusters. In this case, we discretized the age for enhanced visualization into four groups –¡55, 55–65, 65–75, and +75–. Again, differences can be observed comparing both clusters regarding age distribution, mainly in the proportion of the population aged over 75, with 16.01% for Low PSA against 35.8% for High PSA evolution.

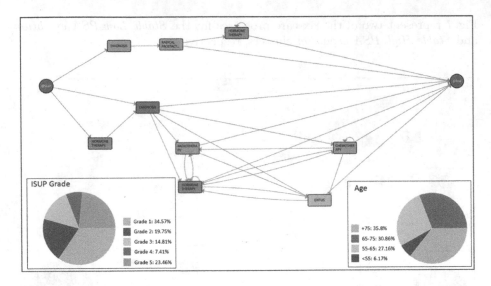

Fig. 7. Care process for Stable High PSA evolution cluster.

4   Discussion and Conclusions

This work presented the results of applying Process Mining techniques to real-world data to obtain prostate cancer processes. In collaborative work, process miners and prostate cancer experts from a hospital developed an IPI characterizing PSA evolution by applying the Dynamic Risk Models framework. Throughout the methodology, Process Mining experts deal with the data, including issues such as availability, quality, and curacy, and infer preliminary sights to the healthcare professionals about the prostate cancer process. Similarly, healthcare professionals analyzed the initial views, identifying processes and considering problems with the data and concrete cases that need special care. Moreover, healthcare professionals identified perceptual questions based on their daily experiences that the Process Mining experts translated into advanced beta views, helping them distinguish between more relevant clinical questions. These novel results might support identifying new information, which might turn to further questions and hypotheses being enriched or generating unique advanced displays that, in the last stage, would help health experts comprehend prostate cancer care processes. The results show differences among the models supporting the use of such techniques to infer clinical knowledge. Future work and next steps should involve health experts in deep analysis to discover new clinical findings and meanings.

Funding. This work was partially funded by the European Union's Horizon 2020 research and innovation program under Grant Agreement No. 875329.

References

1. Bednarik, L., Kovacs, L.: Efficiency analysis of quality threshold clustering algorithms. Prod. Syst. Inf. Eng. **6**, 15–26 (2013)
2. Dyba, T., et al.: The European cancer burden in 2020: incidence and mortality estimates for 40 countries and 25 major cancers. Eur. J. Cancer **157**, 308–347 (2021)
3. Epstein, J.I., Egevad, L., Amin, M.B., Delahunt, B., Srigley, J.R., Humphrey, P.A.: The 2014 international society of urological pathology (ISUP) consensus conference on Gleason grading of prostatic carcinoma. Am. J. Surg. Pathol. **40**(2), 244–252 (2016)
4. Fernandez-Llatas, C.: Interactive Process Mining in Healthcare. Springer, Cham (2021). https://doi.org/10.1007/978-3-030-53993-1
5. Fernández-Llatas, C., Benedi, J.M., García-Gómez, J.M., Traver, V.: Process mining for individualized behavior modeling using wireless tracking in nursing homes. Sensors **13**(11), 15434–15451 (2013)
6. Gandaglia, G., et al.: Structured population-based prostate-specific antigen screening for prostate cancer: the European Association of Urology position in 2019. Eur. Urol. **76**(2), 142–150 (2019)
7. Grüger, J., Bergmann, R., Kazik, Y., Kuhn, M.: Process mining for case acquisition in oncology: a systematic literature review. In: LWDA, pp. 162–173 (2020)
8. Ibanez-Sanchez, G., et al.: Toward value-based healthcare through interactive process mining in emergency rooms: the stroke case. Int. J. Environ. Res. Public Health **16**(10), 1783 (2019)
9. National Cancer Institute: Prostate-specific antigen (PSA) test (2022). https://www.cancer.gov/types/prostate/psa-fact-sheet. Accessed 31 Mar 2023
10. Kurniati, A.P., Johnson, O., Hogg, D., Hall, G.: Process mining in oncology: a literature review. In: 2016 6th International Conference on Information Communication and Management (ICICM), pp. 291–297. IEEE (2016)
11. Litwin, M.S., Tan, H.J.: The diagnosis and treatment of prostate cancer: a review. JAMA **317**(24), 2532–2542 (2017)
12. Mans, R.S., Schonenberg, M., Song, M., van der Aalst, W.M., Bakker, P.J.: Application of process mining in healthcare-a case study in a Dutch hospital. In: Fred, A., Filipe, J., Gamboa, H. (eds.) BIOSTEC 2008. CCIS, vol. 25, pp. 425–438. Springer, Heidelberg (2008). https://doi.org/10.1007/978-3-540-92219-3_32
13. Munoz-Gama, J., et al.: Process mining for healthcare: characteristics and challenges. J. Biomed. Inform. **127**, 103994 (2022)
14. Rojas, E., Munoz-Gama, J., Sepúlveda, M., Capurro, D.: Process mining in healthcare: a literature review. J. Biomed. Inform. **61**, 224–236 (2016)
15. Valero Ramón, Z.: Dynamic risk models for characterising chronic diseases' behaviour using process mining techniques. Ph.D. thesis, Universitat Politècnica de València (2022)
16. Valero-Ramon, Z., et al.: Analytical exploratory tool for healthcare professionals to monitor cancer patients' progress. Front. Oncol. **12**, 1043411 (2022)
17. Valero-Ramon, Z., Fernandez-Llatas, C., Martinez-Millana, A., Traver, V.: A dynamic behavioral approach to nutritional assessment using process mining. In: Proceedings of the 32nd IEEE International Symposium on Computer-Based Medical Systems 2019, pp. 398–404 (2019)
18. Van Der Aalst, W.: Process Mining: Data Science in Action, vol. 2. Springer, Heidelberg (2016). https://doi.org/10.1007/978-3-662-49851-4

19. van der Leest, M., et al.: Head-to-head comparison of transrectal ultrasound-guided prostate biopsy versus multiparametric prostate resonance imaging with subsequent magnetic resonance-guided biopsy in biopsy-naïve men with elevated prostate-specific antigen: a large prospective multicenter clinical study. Eur. Urol. **75**(4), 570–578 (2019). https://doi.org/10.1016/j.eururo.2018.11.023. https://www.sciencedirect.com/science/article/pii/S0302283818308807
20. Vidal, E., Rodríguez, L., Casacuberta, F., García-Varea, I.: Interactive pattern recognition. In: Popescu-Belis, A., Renals, S., Bourlard, H. (eds.) MLMI 2007. LNCS, vol. 4892, pp. 60–71. Springer, Heidelberg (2008). https://doi.org/10.1007/978-3-540-78155-4_6

Author Index

J. M. Juarez et al. (Eds.): XAI-Healthcare 2023/PM4H 2023, CCIS 2020, p. 131, 2024.
https://doi.org/10.1007/978-3-031-54303-6

Printed in the United States
by Baker & Taylor Publisher Services